Animal Camp

Reflections on a Decade of Love,
Hope, and Veganism at Catskill
Animal Sanctuary

KATHY STEVENS

Founder and Director of
Catskill Animal Sanctuary

Kathy Stevens

Skyhorse Publishing Inc.

Skyhorse Publishing books may be purchased in bulk at special
discounts for sales promotion, corporate gifts, fund-raising, or
educational purposes. Special editions can also be created to
specifications. For details, contact the Special Sales Department,
Skyhorse Publishing, 307 West 36th Street, Floor 11, New York,
NY 10018 or info@skyhorsepublishing.com.
www.skyhorsepublishing.com

10 9 8 7 6 5 4 3 2 1

Library of Congress Cataloging-in-Publication Data is available on file.

ISBN: 978-1-62087-566-7

Printed in the United States of America

Dedication

I believe that in all of us is a good and pure knowing that beneath the surface of things, beneath gender and skin color, religion and nationality, social class and sexual preference, and yes, beneath species . . . we are all the same.

This book is dedicated, with love and hope, to that good and pure knowing in you.

Table of Contents

About the Revised Animal Camp

The two comments I hear most frequently are, "Oh, you must love what you do," and, just as frequently, "It must be *so hard*." Honestly, both things are true in equal measure. I can't conjure up work that would be more joyful or fulfilling. Twelve years after we opened our doors, these animals still take my breath away. The contributions Catskill Animal Sanctuary makes toward our vision of a world free from suffering feel good, pure, and purposeful. "Right livelihood" is the term used by Buddhists, I believe. At the same time, the work is relentless and sometimes overwhelming. On both counts, I'm fairly certain my CAS colleagues agree with me. To present just the good stuff might make for a wonderful book or good PR for Catskill Animal Sanctuary, but it is not complete to suggest that the sun shines every moment of every day. Whether it is bumping up against entrenched belief systems that cause unspeakable suffering, or saying goodbye to a beloved animal friend, or constantly feeling the pressure of fundraising to meet an always growing need, this work ain't for the faint of heart.

When Skyhorse Publishing told me they wanted to do a paperback reprint of the original *Animal Camp*, I responded with an enthusiastic, "*Oh no!*" Seriously. I was proud of my first book, *Where the Blind Horse Sings*. But *Animal Camp?* Not so much, to be honest. I felt it was incomplete. Beyond that, the idea of having a three-year-old book represent Catskill Animal Sanctuary in 2013 and beyond felt awful. We've grown *so much* in the last three years! Just as important, America's attitudes towards veganism are quite different than they were just three short years ago, and I hoped to be able to address this exciting shift. If the book was going to be placed out in the world again, at the very least, it needed to be updated.

So I asked if I could do a major revision of the book: We'd keep the strong chapters, cut out the dated ones, and add several new essays that better reflect who we are as an organization in 2013.

To my amazement, Skyhorse said yes. Bless them.

This *Animal Camp* is a collection of personal reflections on the work of Catskill Animal Sanctuary. Grouped thematically rather than chronologically, the essays reveal both the joys and the challenges of our work—the deep satisfaction that comes from removing animals from desperation, for instance, along with the "weight of the world" feeling that sometimes accompanies our efforts to open eyes and hearts to the realities of our food-production system.

Dozens of important books on agribusiness and its treatment of animals have been written in the last fifteen years. Some of the best have been published in the last five. I strongly urge you to read a few. Powerful documentaries have been produced, too: works like *Peaceable Kingdom, Food, Inc., Earthlings, Vegucated*, and more.

A full list of recommendations is included at the back of the book. Yet my favorite work of all about the life of a farm animal is Sy Montgomery's *The Good, Good Pig,* a memoir depicting the long life and daily antics of a pig companion named Christopher P. Hogwood. As much as the many well-researched works have taught me about the lives of pigs, Christopher P. Hogwood helped me *know* pigs.

It is this same void that Catskill Animal Sanctuary and *Animal Camp* seek to fill. There is a dearth of firsthand information about farm animals from folks who know them like you know your kids. These stories, told by people like us who live and work among the animals in joyful environments, encourage a level of understanding of the "beingness" of animals that's otherwise missing. If my musings help you see farm animals in a whole new light, I've done my job. If they go a step further—if they're an exciting invitation to move toward a more compassionate lifestyle, well then . . . the animals have done theirs.

Animal Camp is organized into three sections. The first, titled "What We Do," shares a wide range of CAS experiences with you, from the sublime to the shattering, from actual rescues to tour days, from Chef Linda's vegan cooking class to farewells to beloved animals that somehow feel both crushing and triumphant. This section is an introduction for all the folks who say, "Tell me about Catskill Animal Sanctuary." My only regret is that our wonderful summer day camp, called Camp Kindness, isn't included. It's a unique and life-changing program for children, but the season came and went before I could write that chapter.

"Why It Matters" looks at the stuff that never makes most folks' radar screen—the impact of agribusiness on all of us, for instance. Watch as twenty-pound hens struggle to breathe on a hot day and feel the agony of a pig bound for slaughter. As you delight in the beauty and freedom of four calves once bound for the dinner plate, as I attend a fundraiser for one species at which another one is served for lunch. I also say goodbye to my beloved teacher and friend Rambo in this section, because what happened immediately after he died illustrates what I believe in my bones: In the ways that truly matter, we are all the same. As much as his life mattered, his death mattered, too, in what it taught us and in its implications for how we share what we know to be true about our animal friends.

"When We Smile" is a section of snapshots—essays depicting brief moments in time. It brings us back to joy after the somewhat heavier preceding section. Enter the daily life of Catskill Animal Sanctuary and experience the joy that defines nearly every day. Share my delight in simply observing the antics of "The Underfoot Family," the cast of characters free to roam the entire farm all day long. Given their ability to make their own choices about how to spend their time, they entertain and inspire and teach us. *And* they get in trouble. Arthur the free-range goat winds up in time-out several times a week. You'll laugh, you'll cry, you'll fall in love—and you'll understand why we all consider it a privilege to do this work.

Taken together, the three sections of *Animal Camp* provide a glimpse into a most challenging, unique, wonderful world—a world in which I feel privileged to participate. My fervent hope is that my depiction of this world will challenge you to make the lifestyle changes we must all make, right now, to minimize unspeakable suffering and to heal an ailing planet.

See you at the Sanctuary.

About Catskill Animal Sanctuary

C lose to 2,500 farm animals, victims of neglect, abandonment, tragedy, or the food industry have found safe haven at Catskill Animal Sanctuary since we opened our doors in New York's Hudson Valley in 2001. Many of our animals—particularly the rabbits, goats, potbelly pigs, geese, chickens, and younger horses—find loving permanent homes through our adoption program. Still, at any given time, 200-ish animals, ranging in size from two-pound bantam roosters to 2,000-pound draft horses, call CAS home. That number swells enormously with large emergency rescues. Many of these animals are "lifers"—CAS is their final stop. Among folks looking for animal companions, there's little interest, sadly, in thirty-year-old blind horses, 800-pound pigs, or 1,800-pound cows.

Throughout the years, we've watched in delight as newcomers respond to good food, spacious pasture, deluxe digs, top-shelf medical care, and love in abundance. For some, the healing is immediate; for others, it takes months, even years, to erase dark memories. A crackerjack farm staff complemented by a small army of exceptional volunteers does its best to ensure that each one gets what he needs not merely to survive, but instead *to thrive*. Care is individualized to a degree that plenty of folks would probably consider absurd.

"Heal in your own way, at your own pace, on your own terms," is our unofficial mantra, and it guides how we work with all our animal friends, whether rabbit or rooster, goat or goose. What's permissible for one—something as simple as eye contact—might be threatening for a more timid animal. Both physical and emotional care are highly individualized, from bedding to housing to diet to who lives with—and next to—whom. To be "devoted to their happiness" demands this level of care. And for the caretakers, participating in the transformation of a broken spirit is something all of us consider a profound privilege.

In the midst of it, the most remarkable animals become teachers offering life-changing lessons. In these pages, for instance, you'll meet a Jacob sheep named Rambo, who spent his first year at CAS trying to kill any human who got close to him and his final ten watching out for all living beings at CAS as if we were his flock. To him, *we were*. (His story—our story—begins with *Where the Blind Horse Sings*. I recommend you start with that book, which can be ordered at www.casanctuary.org or on Amazon.com.) There have been many other teachers, too: roosters, turkeys, horses, pigs, cows, and rabbits. If you've had animals in your life, you know how the exceptional ones can rock your world.

Anyone paying this level of attention can't help but notice an obvious truth: farm animals are individuals, with distinct preferences and unique personalities. While there are particular qualities inherent in all chickens or pigs or cows or humans, any ten individual chickens are as unique as any ten humans. Tiny Reef, a two-pound bantam rooster, loves nothing more than to perch on our arms (or our heads) or be carried around all day. He's but one in a long list of exceptionally loving birds. I believe that gentle Jailbird, a rooster rescued from a crystal meth facility in Kansas City, *somehow* understands his role as ambassador for his species as he is passed from the arms of one receptive visitor to the next. We have shy chickens, exuberant chickens, and chickens who are drama queens. And impatient chickens? Of course. A chicken grieving the lost of a friend? Yes. A chicken *trying to tell you something right this exact second if ONLY you'd pay attention?* All the time! You get the point: People who speak in generalizations about chickens reveal how little they know about chickens. The same holds for all farm animals. The vast majority of us knows virtually nothing about the animals we consume.

This brings me, of course, to the second reason our care is so individualized. Not only is it a vital component of the rescue part of our mission; it's an equally vital part of our educational program. Why? It's simple: confident, contented animals are wonderful ambassadors for their less fortunate brethren. Our tours are punctuated by interactions that disarm unsuspecting visitors—unsolicited gestures of tenderness, curiosity, confidence, and affection from animals whom visitors have likely never thought about. When a meat eater holds a cooing chicken, or is accompanied on his tour by goats Arthur and Jimmy or a couple curious turkeys, the wheels begin to turn. When Amos the steer licks a visitor's face, chews his shirt, and uses his leg as a scratching post, a hamburger is no longer "a hamburger." These are moments of rare and powerful clarity.

When epiphanies like this happen, our signature programs—from Compassionate Cuisine cooking classes to Camp Kindness for kids to day-long events attended by hundreds—support good people interested in aligning their lifestyle with their values.

It's simple, folks: each single person who adopts a vegan diet saves as many as one hundred animals *a year*, so as a non-profit whose mission is to save farm animals, our greatest impact lies in encouraging and assisting people in adopting a diet that either contains far fewer of those animals or excludes them altogether. In the process, that person also gets healthier and treads so much more lightly on our frail and desperate planet. Best of all, perhaps, she gains the peace of mind that her circle of compassion extends to *all* living creatures. See Appendix One, "All the Right Reasons," for a few of the ways that a vegan diet is truly what Alicia Silverstone (and so many others) calls "The Kind Diet" in her cookbook of the same name.

Throughout the last few decades, we humans have sat by as food production has been concentrated among a handful of multi-billion dollar companies: Tyson, JBS, Dean Foods, Cargill, and others. These companies have plenty of friends in high places. Their directors sit on the boards of our largest financial companies, our giant energy and pharmaceutical companies, our largest agricultural universities. They are friends with powerful lobbyists and politicians. Their influence on food policy, environmental policy, animal welfare standards, and more is titanic. Power and profits are what matter to them: damn the people who eat the toxic food. Damn the earth being used up past her breaking point. The animals grown to feed us? Feed 'em fast, feed 'em cheap. Their mechanized system of growing animals to turn them into food for humans subjects billions every year to a level of deprivation and suffering that no human being with a shred of compassion would wish upon another living thing. The system is an abomination.

Consider, for instance, the devastating impact of animal agriculture on our planet. With a growing human population demanding more animal products, there is an accompanying demand for more water, more land, more feed for the animals, more fertilizer and pesticides and antibiotics (all toxic, all entering our soil and water), more fuel, more electricity, more waste disposal capacity, and on and on. *The planet can't get bigger to accommodate this ever-increasing demand. The planet is coming apart,* a fact that has been presented by dozens of organizations from the United Nations to WorldWatch Institute and borne out by environmental degradation, human illness and cancer rates, and the increasing global climate instability of the last few years. Look at the year 2012 in the United States alone: Irene. Sandy. Lee. A relentless drought that impacted much of the country and resulted in the culling of animal herds, hay and grain shortages, and price hikes. Near the end of the year, a National Public Radio commentator remarked that 2012 was the year that made global warming real. He was right. It did. We're paying attention.

In all the science around global warming, two statistics are especially useful: 1) Raising animals to feed humans is the single largest contributor to greenhouse gases, accounting for 18 percent of all emissions that are cooking our planet (one study believes that the figure is more like 50 percent—see the "Air" section of "All the Right Reasons" at the back of the book), and 2) with the same amount of natural resources that it takes to feed one meat eater, one can feed sixteen to twenty vegans. It is pretty clear: our greatest hope of slowing, and eventually reversing, global warming is a paradigm shift to a plant-based diet.

Whoa there, girlie! With a challenge of this magnitude, what difference can a small non-profit like Catskill Animal Sanctuary make? Plenty, as we see it. On the simplest level, almost 2,500 animals—blind sheep, horses, cattle, and ducks; rabbits found in sealed Tupperware containers; a horse who survived a bullet wound to the head; hundreds of victims of mentally ill hoarders and so many others—have come to a place that, to them, hopefully feels like heaven. If one believes that each life counts, then places like Catskill Animal Sanctuary matter.

Regarding the shift to veganism, no matter your motive— minimizing suffering, helping humanity heal the planet, or healing your own body worn out by a meat- and dairy-centric diet— Catskill Animal Sanctuary is here to help. Whether you want to begin by reducing your consumption, or are one of those rare beings like our friend Michael Graff, who went vegan overnight after hanging out with Dozer the steer, our programming at Catskill Animal Sanctuary is here for you the moment you say, "I need to change my diet." From programs such as Camp Kindness, a day camp for children, to Compassionate Cuisine (register for classes early: they sell out fast!), to our special events, to our website loaded with recipes and resources, we will be both your coach and your cheerleader as soon as you're ready to take the first step.

So welcome, friends, to Catskill Animal Sanctuary. Laugh. Cry. Fall in love. Go vegan.

What We Do

People often ask us, "How do you get your animals?" I used to think it such an odd question, until I realized that most people are only familiar with dog and cat shelters, in which the stream of people surrendering their dogs and cats never stops. Sadly, there is often little room at sanctuaries for the animals whom well-meaning people can no longer care for. Unless the need is truly urgent—for example, when a caretaker has died—these animals are usually placed on our waiting list. During rare times when space is available, we happily accept animals from the list. Mostly, though, we don't have space, because "emergency rescue" truly is *emergency* rescue–our animals often come a dozen at a time, even scores at a time. They come from animal hoarders or from infinite variations of situations one would call "bizarre" or "disturbing."

We are often, of course, asked to help with the physical rescue itself. Reasons vary, but they boil down to fewer people knowing how to catch, handle, safely load, or haul farm animals. Even when there is a cruelty case, law enforcement relies on us to remove the animals. My guess is that most readers have been in a car with a dog or a cat; not so much a goose, a pig, or a turkey. Few have loaded a terrified horse or cow onto a trailer. Many of us have joined friends in a search for a missing dog, but few have traipsed up a mountaintop and tried to catch a herd of injured feral goats. Come along with us and learn more than you ever knew existed about what "farm animal rescue" entails.

In "Welcome to Catskill Animal Sanctuary," you'll accompany me on a tour, meet some of our animal friends, and consider their lives from a perspective that perhaps you never have. After you've fallen in love, tiptoe into Vegan 101 with me, where Chef Linda Soper-Kolton is helping the veg-curious create three quick, simple, tasty entrées. She shares her recipes from the class, as well! We hope you'll try them and let us know what you think, and we hope that "sitting in" provides the encouragement you need to begin your own journey toward a healthier, kinder diet.

So there you have it: a peek into life on the farm and another into two of our many educational programs. We work hard to mend broken bodies and spirits. We work hard to strike a delicate balance between honesty about what animals endure at our hands and respect for each person's journey; we work hard to empower you to make positive, healthful changes. I hope you'll let us know how we do.

Just Another Day at CAS

Eight of us are climbing a narrow path up a Pennsylvania mountainside. Granted, this is Pennsylvania . . . we're not talking Kilimanjaro. Still, the climb is rocky and uneven, and our hands are full—we will likely be tired before the work even begins.

Troy has a large dog crate balanced on his shoulder; Walt carries backpacks filled with rope, sheets, lead rope, and a first-aid kit. Volunteer Sharon Ackerman, an acupuncturist, carries water for the troops; the rest of us pass heavy crates between us. I carry one atop my head for a while, then pass it on to volunteer Vanessa VanNoy, who's taking a day off from yoga instruction. (Vanessa loves adventures like this. She and her husband once flew to Kansas City, rented a Budget truck, and drove forty-two chickens back to CAS after they were rescued from a crystal meth lab.) Lorraine, meanwhile, doesn't even work at the Sanctuary on Wednesdays, but here she is, climbing with us nonetheless, as is Friday volunteer Debbie Wierum. We are going to rescue nine feral goats. At least one is injured and unable to walk. Several are pregnant. All are emaciated.

"I see one of the black ones," says Gray Dawson, who is just ahead of me. His wife, Melissa, has left the group and gone ahead of us. It was Gray and Melissa who called us about the animals. Melissa has seen the goats five days in a row now at the top of her friend's vast mountainside property. Her friend can't fathom who they belonged to. "No one around here raises goats," he tells her. Gray and Melissa are concerned that the goats won't survive, hence their call to CAS. If their descriptions of the goats are accurate, they're right. It sounds like there's not nearly enough available food to sustain them, first of all; secondly, they're prime and easy targets for coyotes.

Per our instructions, the couple has been taking small amounts of grain to the animals, hoping to earn their trust. When we've gone half an hour up the mountain, Gray and Melissa instruct us to hang back as they continue on to locate the goats. So we take our positions in a pod, sitting very still, until sure enough, two pygmies come bounding down the mountain. Walt moves forward. In an instant, the two little goats have lead ropes around their necks and are munching happily on grain. We lure them into the biggest crate, knowing we'll need all available humans to safely catch the seven remaining goats

"We'll never get them," Melissa says dejectedly.

It's the pygmies who have come to her several days in a row. Of those remaining, only one has come cautiously within her reach in preceding days. For the others, even though their emaciation is extreme, fear has consistently won out over hunger.

I am far more confident than Melissa, and though we don't talk about it, I suspect the rest of the CAS crew is, too. It's not like we've had a lot of experience rescuing abandoned goats from mountainsides. But we have taken in hundreds of animals who were originally (though usually not for long) terrified of us. We know how frightened animals move. We know goats can spring and leap with remarkable speed. Our bodies know how to match their movements.

The injured goat and her very pregnant friend are huddled at the edge of a vertical rock face. There's a thirty-foot drop to the craggy, boulder-filled forest floor. If a terrified goat leaps, she'll die; if a human slips, ditto.

"Should I try to get behind them to drive them down the hill?" Melissa asks.

"Yes," I say, and as she does so, the rest of us very, *very* slowly encircle the two frightened animals.

A boulder juts upward between Lorraine and me—a boulder that a frightened goat will very likely use as a springboard to leap over our heads and out to freedom.

"Kathy," Lorraine says. I turn; she's found a long, large branch. We stretch to hold it between us, hoping an additional visual barrier will dissuade a fleeing goat.

We close in. We are all crouching, we are all quiet, we all utter soft words of encouragement. We are fifteen feet from the goats. Then we are twelve; then we are six. A few more feet and we can dive and catch them. But a goat is flying through the gap between Lorraine and me. I leap for her and hug her to my body. We collide in a heap to the ground and are sliding downhill toward the cliff. Instantly, Walt and Troy are there, wrapping the frightened animal in ropes and a sheet to keep her both safe and contained. I sit up.

"Good tackle!" a voice from the left says.

"Thanks," I say, quickly dusting myself off.

Over the next hour, we continue this methodical work until just one goat remains. She's below us, backed into a corner below the rock face.

Troy scales down a steep slope, followed by Debbie. I can't see the goat.

"If all of us come down, can we trap her?" I ask.

"I think so," comes Troy's measured response.

One by one, we lower ourselves into the small cavern. This time, the goat can only move forward; the rock wall behind her is sheer and solid and completely vertical. We form a line and inch forward, our hands outstretched to close in the gaps. The goat, rail thin but pregnant, turns quickly in every direction, desperate to find a way through. She leaps, but Troy leaps faster. They fall together and then Walt is there, safely harnessing her and wrapping her terrified, weakened body in a sheet.

It's clear we'll need to carry her out . . . and we do.

⁓

Three hours after we arrive, we load nine goats safely into a warm rented cargo van bedded with straw and head north to Catskill Animal Sanctuary. The goats will live in a large, hilly pasture custom-made for goats. The back half is filled with trees and boulders, much like the forest they came from; the front is a grassy meadow. Their babies will not fall prey to hungry fox or coyote. In this new home, they will have plenty to eat, warm shelter in the winter, and, if they choose it, plenty of love from always-willing humans. If not, there's ample room for them to continue living virtually the same way that they were living, only with plenty of added amenities, including safety.

It's six o'clock when we return to the farm, but the rest of the staff have all waited for us. We settle the goats into their deeply bedded stall; fresh hay and water are piled in opposite corners. Tomorrow, we'll separate out the pregnant moms, whose udders suggest that kids could arrive at any time. We're tired, but smiling. It's just another day at Catskill Animal Sanctuary.

The Little Horse That Could

"He's too weak to walk," Stephanie Fitzpatrick of the Dutchess County SPCA told me before she pulled into the driveway with our latest rescue. She wasn't exaggerating—little Andy collapsed when he got off the trailer and then fell again in the driveway. Looking at him now, this little white horse with hundreds of brown spots, I can't believe he is alive. He's nothing but bone. But he *is* alive . . . and that means we've got work to do.

Andy's owner runs a "nurse mare" operation that rents out mares for their milk. Unless they're to be used for breeding, male horses serve no purpose. Like male chicks in the egg industry, male goats and cows in the dairy industry (prior to the production of veal), or male horses in the industry that produces estrogen replacement drugs from the urine of pregnant mares, they are considered by-products and are discarded by any means the business owners wish. Baby chickens, for instance, are suffocated, crushed, gassed, or ground up by the millions by the egg industry; the industry has no use for the birds who don't produce their product.

Andy, too, was "useless," and he was dying from starvation.

Coming into work the morning after Andy's arrival, Alex burst into tears when he saw our new friend. "How could *any* living being do this to another living being?" he asked.

It's everyone's reaction. As a three-year-old Appaloosa stallion around fourteen hands, Andy should weigh eight hundred pounds or so, but at most weighs half that. If you're of normal weight, picture yourself with half your weight missing: you'd be skin draped over skeleton. That's Andy.

"I don't know how he's alive, quite frankly," said veterinarian Heather O'Leary. She labeled him "below 1" on the Henneke Scale, a system used by veterinarians to represent horses' body conditions, with 1 being the lowest possible number on a 9-point scale. A Henneke 1 depicts a horse in danger of dying. In many previous cruelty cases involving more than a hundred horses total, we've had only seven horses considered by examining vets to be Henneke 1; none were nearly as debilitated as Andy. Surely nothing but will has kept him breathing.

Keefe has just arrived. Trying to prepare her for what she's about to see, I have taken her hand and led her to the stall. We stand in front of his stall looking in at the little horse, who looks part Dalmatian, part deer, part horse. She cries, too, managing only to say, "He's so frail."

"I know," I respond. "This one's going to be a challenge."

Andy's legs are pencils; his rump higher than his shoulders due to absurdly long rear legs. He strains to poop, but passes water instead of manure. Later, his urine is the color of dark chocolate. We notice a thick pus-like substance running down his back legs. His rectum has prolapsed; it hangs outside his body, a raw and painful mass. We wrap his tail in an ace bandage to prevent further irritation and infection.

"Do you think he'll make it?" we're asked a bazillion times a day.

"He made it here alive," I say to everyone who asks. "This is easier than what he's been through."

Of course, the staff knows that Andy's recovery is uncertain. In fact, he could be in organ failure. But Andy is one of those animals with whom everyone has fallen instantly, hopelessly in love. We love him for his courage, for the light and brightness in his eyes in spite of what he's been through: chronic starvation for probably his entire life. He is going to fight; we will fight with him. Part of that fight is surrounding him with hope and positive energy. So volunteers and visitors hear that Andy is doing beautifully, even when he's struggling.

A few days after Andy's arrival, I enter his stall and sit with him for a long while. He is lying down. His sides rise and fall hard—even the act of breathing is difficult. "*You did it, little man! You're alive, little trooper. Good job!*" I whisper to him as he rests. A lesser spirit would have given up long before now. He sits up and looks at me without malice. No anger, just exhaustion and quiet determination. He nibbles my knee. "Hey, silly boy," I say to him. When he decides to stand, I back away to give him room. Left front leg stretches out, then right. Andy struggles mightily; his entire body trembles.

"You can do it, Andy!" I encourage him.

Andy is shaking violently, but he has made it. He is standing and I am weeping; it's the first time I've seen the heroic effort required simply to stand. ("It's like watching a baby stand for the first time," Alex says.) Andy leans against the stall wall for support. I lean into him with my full weight, my back just behind his shoulder, thinking that the extra support might be enough to help him last a few minutes longer. Even if it isn't, I want to convey that we're with him in this. His sides heave from the exertion; his tiny legs tremble. After no more than two minutes, it's time for me to step aside.

"Okay, *Andrew*! What a great job!" I praise him. Andy folds his front legs under him and collapses with a thud.

We've triple bedded Andy's stall with three heaping wheel-barrows of shavings. We make jokes about the "Princess and the Pea." But even with this generous bed, lined with rubber mats, Andy has large raw abrasions on his hips from where the bones rub the floor. We treat them and add more shavings.

I lie down in front of Andy's head; the two of us form a T. I rub his cheek, his neck, and his forehead, and for a few moments, he nibbles my shirt. Within minutes, he's sound asleep. (He sleeps for most of the day, of course. In addition to a medically-supervised diet and a growing fan club, quiet, comfortable sleep is what he needs to heal.)

Andy's owner has been arrested and charged with cruelty. While her fate plays out in the courts, we will do all we can to heal this lovely boy, whose spirit has captivated us. After just four days, he already wobbles on unsteady legs to his front wall, leans out, nibbles a cheek, pulls a hat off an unsuspecting head. Too weak to muster a full whinny, he calls out a pitiful "hello"—all breath, no sound—to everyone he sees. Yep, he's a trooper, all right. And we're rooting for him.

～◯

October 17: It's been a month since Andy's arrival. He's still a mess, but at least he's not contagious. That's a piece of very good news. We put away the foot baths. It feels good to touch him with my skin rather than with a gloved hand. It feels, as Keefe says, "like our new baby is finally home from the hospital."

Andy is still a bag of bones, still extremely wobbly on his feet. He's gained perhaps forty pounds but is still so skeletal that there's no point trying to use a weight tape, a measuring tape used to estimate horses' weights that is supposedly accurate within twenty-five pounds. Veterinarian Heather O'Leary says his condition will skew the results.

One has to use extreme caution when putting weight on a debilitated horse. They need to be fed small frequent meals of hay only, gradually increasing the amount, decreasing the frequency, and eventually adding small quantities of grain as time passes. Andy's case is especially sensitive—his starvation was chronic; his organs so stressed and compromised. Like so many who've come before him, Andy is on a special refeeding program. Yet despite the measured discipline of Andy's weight-gain regimen, he's getting far more food than he's ever received. Nearly as soon as he finishes a flake of hay, another one is on its way—an alfalfa mix; lots of energy, lots of calories. He eats with relish. Soon, we'll add grain to his diet.

Just as important for Andy's healing, however, is the attention he's receiving: the kisses, the gentle words, the grooming, the encouragement. It's rare to see Andy alone. Normally, one or more of his sizable fan club are just outside his stall, either stroking his face or allowing him to chew their coats, their hats, their hair. He *loves* the attention! Keefe is among the many who love to groom his tiny frame. In fact, she says that even as he playfully bites her arm, she's happy to see "the little punk" full of life and ready to play.

If Andy could speak our language, he'd probably tell us delightedly that he's the first CAS resident to receive a daily sweeping. Mind you, I don't make a practice of sweeping horses' entire bodies. But one day, when I simply *couldn't* sweep the aisle outside his stall because he kept on grabbing either my shirt or the broom handle, I turned his playfulness into a game.

"Okay, Andy, you want the broom? You can have the broom!" I said, and then brushed his entire face from ear to muzzle. Andy grabbed on, and we played gentle tug of war until a perfectly good broom was about to be destroyed, at which point I dropped the handle, went to him and opened his mouth, forcing him to let go. I entered the stall, and I swept Andy's neck, his back, his rump. Andy stretched his neck out as far as he could; he liked this game!

This work—if you can call it work—is one of the greatest joys of what we do. Any of us can tell you what a privilege it is to participate in the healing of a broken animal—to say with every word, touch, gesture, "You're safe, sweet one. You made it." Andy is the most exuberant spirit I've ever known and thus makes our jobs that much more rewarding. But whether joyful and outgoing or fearful and reserved, all hearts yearn to sing. What a humbling and extraordinary gift it is to be one among those who offer that opportunity.

Andy is definitely standing more solidly on his pitiful pogo-stick legs. His manure is generally solid; his rectum is staying inside his body. And it could be wishful thinking, but I don't think so: I swear his whinny is getting stronger. It won't be long before Andy is ready for his first real adventure: a walk outside to explore his new world at Catskill Animal Sanctuary—a far cry from the nightmare that cruelty investigators found him in.

~⌒

One hundred and twelve days—that's how long it's been since Andy arrived. That's how long it has taken Team Andy, comprised of veterinarian Heather O'Leary, master farrier Corey Hedderman, former Animal Care Director Walt Batycki, and dozens of others offering love, encouragement, and time, to bring him to this day. One hundred and twelve days. But we've arrived, and he's ready. Andy is going outside.

"Come on, little guy," Walt says to him as he attempts to put the magenta nylon halter over Andy's head. It's a real challenge, of course: Andy wants to chew the halter. Andy wants to chew *everything*. He grabs the halter quickly, again and again. Finally, I hold his head while Walt slides the halter on. Though we've been working on this, it's clear we have a long way to go. (Andy is now four years old—a young adult—but everything about him is still a baby. I don't know whether it's simply who he is, or whether, deprived of a mother and a herd of horses to socialize with, Andy was never taught which behaviors were acceptable and which ones were off–limits. Certainly no human ever taught him. It doesn't matter anyway. His mischievous nature endears him to everyone he meets.)

"Andy, we're going to take a walk!" I exclaim. Walt clips on the matching magenta lead rope.

"You ready, fella?" he asks. A gaggle of staff and volunteers has come in to witness Andy's big moment.

"I can't believe he made it," Melissa, a volunteer, offers. "I can't believe he made it to this day."

Tears fill my eyes and I pause for a moment. "This is a good one, guys," I say to the group. "Congratulations."

"Aw, shucks," Walt says. "It's been a blast working with an animal who tries to eat my face every single time I go in the stall."

It has been a blast . . . for all of us. Indeed, it has been an honor.

"You ready, fella?" Walt asks again.

"Here you go, boy," I say as I open his stall door wide. We stand, our arms around each other, and watch in awe as the Little

Horse That Could steps out of his stall. All Andy does on this first day is exit the barn, circle it, and come back into his stall, walking perhaps a total of three hundred feet. But every single one of us witnessing this moment recognizes it for what it is: a miracle.

I enter the barn from the side near my house, and as I glance down the aisle, Andy is doing his "Oooh, it's recess! I get to go play!" dance as Keefe leads him from his stall.

"Hold on, Mister!" she says, and as I approach them she adds, "You know, somebody needs to tell this boy he was nearly dead nine months ago. Apparently he's forgotten."

I turn to walk outside with them. It's not that I think Keefe will have trouble; she can handle an agitated horse just fine. It's that I want to witness what's about to happen, for Andy's first moments in the pasture are always worth the price of admission.

Andy shares a field directly behind the barn with five other horses: geldings Noah and Bowie, and mares Callie, Hazelnut, and Crystal. These five stay in the field round-the-clock. We call them the Kind Group. There's not an attitude in the bunch. Bowie and Noah— each of whom had his own stall in the barn until their physical and psychological healing was complete and whose behavior suggested they'd be happier as outside horses—are the newest additions.

I open the gate and watch as Andy chews on the sleeve of Keefe's jacket and does a little jogging in place. He's like this each morning: antsy. She leads him in and releases him, saying, "Go get your buddy, Andy!" In just two strides, Andy is in a full-out gallop, charging across the hill to his best friend, Bowie, who lifts his head as Andy approaches. As Keefe returns to the barn and her long list of chores, I follow the spotted imp.

Andy's pal Bowie, a lovely quarter horse, arrived just four months ago. He was two hundred pounds underweight and cowered when anyone entered his stall. While he was far mellower than Andy, what the two horses shared was a strong desire to get past their history. That plus youth: while the older mares found Andy's insistent personality annoying, and Saint Noah tolerated him, Bowie, just five years old himself, found in Andy the perfect playmate. Their friendship was nearly instantaneous.

Andy has found his pal, and for a long moment, the two young boys stand head to head, breathing each other in. Then, without warning, Andy does a funny little scoot backwards, then a quick scurry around to the right until he's at Bowie's rump. He nips his friend playfully on the fanny, and the games begin. For the next twenty minutes, they play as young foals play—cavorting and kicking, bucking and rearing, charging through the field. Bowie stops short and Andy stumbles. No matter. He's back up in a flash, grabbing Bowie's tail with his teeth.

One year after the Little Horse That Could arrived at Catskill Animal Sanctuary, he still looks part horse, part gazelle—the likely result of inbreeding and chronic, extreme malnourishment. His back legs are a good four inches too long; his hocks jar violently to the side each time he takes a step. He'll never win a beauty contest. But that doesn't matter. What *does* matter is that Andy recovered. Class clown. Nudge. Annoying little brother. Victor. Teacher. Hero. He's the little horse that could . . . and he did.

<hr>

And then one day, Deborah Smith-Fino showed up.

"I'd like to adopt two horses," she said. "One for me, and one for my husband."

Deborah, in returning to the farm several times, fell in love with Bowie; her husband with the alpha mare Athena. We certainly understood their choices! Young, striking Bowie—with his bald face, one pale blue eye and the other half brown, half blue—was another survivor of horrific abuse, and when Deborah appeared, Bowie was just learning to trust us. Deborah was smitten. Strong, confident, outgoing Athena was a no-brainer for Deborah's husband Eric. The couple's adoption application was approved pending completion of their barn, and they got to work to finish it.

Fast-forward to a conversation between Deborah and me in which I mentioned that Bowie and Andy were best friends who grazed nose to nose, charged around the field together like two young boys should, rested chins on each other's rumps. Given their individual histories, I was loath to split them. Noah, a far more mature horse, would bond easily with anyone; he actually, after a time, found Andy rather obnoxious. But youngsters Andy and Bowie were attached at the hip.

A few days after that conversation, Deborah appeared in the barn. She watched as a tour stopped in front of Andy's stall and Andy tried to chew everyone who came within reach. She watched as I swept Andy with a broom—he still loved the game. She heard stories about how Andy fell sound asleep with his head in my lap and how he allowed me to stretch out fully on top of his prone body.

When the tour continued down the aisle, Deborah stayed behind with Andy.

Two hours later, I dropped off my final group at the Welcome Hut, then returned to the barn to load up the truck for afternoon feed. To my surprise, Deborah was still with Andy.

"I want him," she said softly as I walked back into the barn.

"I don't want two animals who have been through so much to lose each other." And she said it again *after* she read through Andy's medical records—all thirty-some pages of them, including the prolapsed rectum, the liquid manure, his struggles to stand, the farrier work when he was far too weak to stand on three legs, the vet's description of his deformed body and how it might impact his health.

A few days later, our veteran (and vigilant) home inspectors Walter McGrath and Charlotte Mollo gave an enthusiastic "thumbs up" after their follow-up inspection of Deborah and Eric's newly completed barn.

The next Sunday afternoon, our hauler Corinne wound down our driveway with her fancy white trailer. Athena went on first, followed by Bowie. Andy was next. As Allen walked him up the aisle toward us, my tears began to fall.

"Do you mind if I load him?" I asked Allen.

"No, take him," Allen said with generosity.

I took him, our Little Horse That Could, and led him up to the trailer. Andy hesitated. Corinne was inside; I handed the lead rope to her and I stood right next to Andy, my hand on his neck.

I'll never forget what happened next. Andy looked directly at me, and inside my head, I heard the words, "Are you sure I'm going to a good place?" as surely as if Andy had actually spoken them.

He was looking hard at me. He was shaking. He wanted an answer.

"Yes, my boy. You're going to a good, good place. I promise." I said to him. And he believed me. Andy took a moment to work up his courage, then stepped onto the trailer and stood right next to his best friend Bowie. We closed and locked the rear door, and they were off.

Thank you, Deborah.

And thanks especially to you, Andy. You're in our hearts, little man.

Welcome to Catskill Animal Sanctuary!

"**W**elcome!" I greet a cluster of visitors underneath the willow tree where folks gather to wait for the next tour. Rebecca has taken the 10 a.m. tour, Alex the 10:30. I'm here for the 11 a.m., and on this exquisite June day, it's a biggie.

"*CAN WE SEE COWS?*" a little boy bellows, hopping in place as he does so. His name is Dylan.

"He's been asking all morning," his mom explains. "His favorite book is *Click, Clack, Moo!*"

Dylan, of course, is in luck. I'm about to mention our three sweet steers Tucker, Amos, and Jesse, all of whom are tour favorites, but Mom pipes in, "He already knows them all! We've watched the video of the cow playing in the gravel pile about thirty times."

The group is loaded with children. Other than a young, hip lesbian couple and an outgoing couple in their sixties, the remainder are families: the solo mom with Dylan the cowboy, plus six parents and too many kids to count.

I introduce myself, ask everyone to do the same, and fast forward through the history and mission of CAS, apologizing to the grown-ups who are here without children. "We want you to enjoy these delightful animals, and we hope you'll ask questions about why we believe it's vital for all of us to shift to a plant-based diet. But when most of our guests are five years old, getting to the learning part gets a little tricky." In groups with this particular composition, there's sometimes more cow kissing than information sharing.

"That's okay," a woman of about forty says. "I'm up for a good cow kiss."

Catskill Animal Sanctuary is open for tours every weekend from April through October. Tours are ninety minutes, and they begin every half hour starting at 10 a.m. For a lot of reasons, there is no "standard" tour. Weather, time of day, what animals are near the fence and who is grazing in the distance, which of The Underfoots—our free-rangers—are around at a given moment, all impact the tour experience. So does the size and composition of the groups. We want our guests to have personal, heart-opening interactions with the animals, and we want to provide a deeper understanding of issues related to the consumption of animal products. One-size-fits-no-one. We have information to share and remarkable animals to meet, but no script, allowing each of our six or seven tour guides to tailor each tour to each group. As tour guide extraordinaire Rebecca Moore explains, "One tour can be a total joy-fest, and the next can be like a support group, with *real* hand holding as people absorb information they can't believe no one told them before."

Beyond this level of individualizing, each of us has our own style and our own relationships with the animals, so the "personality" of a tour varies according to the person leading it. Alex, for instance, likes to bring the cows to the fence to show people their mass, and to point out that dairy cows are now one-third larger than they were just thirty years ago. He also likes for people to experience the cows' gentle and affectionate interactions with him. Michelle is madly in love with Amelia the pig, so her tours always include a stop to visit "her girl," who comes dashing across the field at the sound of Michelle's voice.

Like Alex, one of my primary goals is for folks to witness the depth of relationship I have not only with some of our cows, but also with our pigs and chickens. I want them to see the animals we eat as individuals with distinct personalities and preferences; I want them to know, because they witness it, that it is possible for a pig or a cow or a chicken to love a human being and to express that love so clearly that no one will question it. I hope that these are more than "feel good" moments; I hope that guests' views of these animals will be so profoundly changed that they feel compelled to change their eating habits. Yep; on tour days, that's my #1 goal.

On tours like this one, which includes three children under four, we crawl along, meeting fewer animals than we would on an adult tour; I hope that the personal and unhurried interactions make the experience memorable for all, regardless of age.

We begin at one of two small pig barns, where Nadine, who is really just love in an 800-pound pink package, grunts happily as several visitors, mostly little ones, stretch out across her prone body as if she were a sofa. She's one of seven farm pigs grouped into two "families" who live on separate sides of the spacious, airy barn.

Jangles, currently our only senior farm pig, bunks with the potbellies in a different barn.

"I call it 'taking a nap,'" I explain as I stretch myself over the length of her body. "And I would never suggest it if Nadine didn't enjoy it."

"How would you know if she weren't enjoying it?" one of the dads asks. His tone is skeptical.

"Because an annoyed pig growls like a bear," I say. "It's loud and intense and immediate. A happy pig does what she's doing right now. That grunt is her contented sound."

I offer a quick lesson on the most obvious pig sounds: the "oink" that signals contentment (it's used for everything from "good morning" to "thank you for this fabulous food" to "isn't it a beautiful day" to "*I love this straw bed so much!*") to the growl used when they're annoyed, to a disconcertingly human scream of fear, to their wonderful laugh: *ruf-ruf-ruf!* Yes. Pigs laugh. They plot and plan. They dream. My favorite pig sound is the one they make when they're feeling close to someone. I call it their "I love you sound"—a soft, high-pitched coo.

Across the aisle, either Reggie or Roscoe growls as if on cue.

"What was that about?" someone asks. I look over; the youngster Amelia is rooting in the straw as the older pigs nap.

"She annoys them," I explain. "She's pushy, even for a pig, and she has a thousand times more energy than they do. They were telling her to leave them alone."

For a few minutes, most of us turn our attention to Amelia, resident hellion, and her barn mates Roscoe, Reggie, and Piggerty. When I turn back around to Nadine, a young woman is kneeling in front of her, crying. "What is wrong with us?" is all she says.

(Our hope, of course, is that everyone leaves the CAS experience asking that question and vowing to make changes in their personal lives to improve the lives of animals.)

I speak about some universal pig truths: their intelligence, physical strength, emotional range, willfulness, and sensitivity. Beyond those givens, however, ten pigs are as individual as ten humans, and I point to our pigs as examples: the moody and hypersensitive Franklin, the bull-in-a-china shop Amelia, the serene Nadine, the somber but gentle Roscoe.

"How do you know these things?" the same dad asks.

"How do you know the qualities of your children, or your best friend, or your dog? When you spend time with someone, when you pay attention, you know him or her. Species is irrelevant—we're all individuals," is the gist of what I say.

On the way out, we stop at our model of a gestation crate. It's made of wood, not metal, but it is the same dimensions of the cages in which pregnant pigs spend nearly all their lives, until exhausted, used up, stressed, and defeated, they are slaughtered. They've never felt the sun on their faces, never breathed fresh air, never rooted in the dirt or made their nests the way our pigs do with such care, both pushing the straw with their snout and picking it up by the mouthful until their beds are exactly how they want them. Instead, they have spent their entire lives in filthy warehouses on concrete and metal slab floors, often covered with untreated sores, breathing ammonia-laden air.

"What kind of society allows this?" I ask the group. "This is what we support with our diets." Because of the number of young children, I leave out the graphic details, leave out the barbaric farrowing crate that reduces mother pigs to milk-producing machines, the way pigs are grown, the way they are loaded into transport trucks, the horrors of transport, and the slaughter process itself. Instead, I recommend Jonathan Saffran-Foer's *Eating Animals* and Melanie Joy's *Why We Love Dogs, Eat Pigs, and Wear Cows.* In the inimitable style made famous in his novels *Everything is Illuminated* and *Extremely Loud and Incredibly Close,* Saffran-Foer depicts in vivid detail the lives, and deaths, of food animals. Joy's book discusses "carnism," the "invisible belief system" that enables humans to view some animals as food, others as pets. Both are life-altering reads.

⁓

Just down the lane, Jesse the steer is even more affectionate than usual, eager to lick a few unsuspecting visitors with his scratchy tongue. His friend Amos is lying near the fence, his great horns pointing skyward. As I climb the fence, I tell their story: They were among more than two hundred animals rescued by a coalition of animal groups when Catskill Game Farm closed in 2006; the boys have been with us since that time. Youngsters when they arrived, they're now seven years old.

When I look out, I notice Dylan, frozen in place beside his Mom. His eyes are huge.

"He's never seen a *real* cow up close like this," she explains.

"There's a slight chance we might be able to do something pretty special," I say to her. "Lemme see what's going on in here."

I discuss the importance of being super careful around animals with horns. "These guys could do some serious damage just flicking their heads to shoo a fly away," I explain.

Beef and dairy operations minimize risk by removing horns, of course. A variety of methods, all barbaric, are used. All cause excruciating pain. Caustic pastes, hot irons, saws, and knives are a few of the methods. Guillotine dehorners are also popular. These tools look like tree pruners and, according to the industry website dehorning.com, carry risk of exposed sinus, infection, and blood loss, as well as setbacks in overall health. The animals do not receive anesthesia or painkillers during this standard agricultural procedure, even though the industry acknowledges the associated pain and posts training videos, marked with the warning "graphic content," in which cattle bellow and writhe in terror and agony.

Obviously, CAS would never consider this excruciating process. Instead, we know each individual animal and are smart and careful around the challenging ones—generally young, exuberant steers always eager to play.

I step into their field, walk slowly toward Amos and kiss him on the forehead just below his impressive horns.

"Here's the really cool thing about knowing your animal friends. You get to do things like this," I say as I stretch one leg over Amos's massive back, then lay my chest on his neck and give him a hug.

I think this is the moment when young Dylan stops breathing. He whispers urgently to his mom. I imagine it's something like, "I HAVE TO DO THAT RIGHT NOW MAMA!! I HAAAAVE TO *RIGHT NOW* OR MY FACE WILL FALL OFF!" I watch her shaking her head "no" as she looks toward me. Yet based on where Jesse is standing and how Amos is positioned, it is actually completely safe to grant Dylan's wish. Amos is a placid, affectionate animal accustomed to having me sit on him for a good long hug. However, this is only the second time I've ever brought a child in with the great beasts. After the death of our gentle giant Babe in 2012, that honor was passed on to Tucker, who has no horns. But a few seconds later, Dylan is in the field with me, a remarkably trusting mother having allowed the visit, his tiny body draped over a contented steer.

For the rest of the tour, Dylan is silent. I suspect that every cell of his tiny body is trying to absorb what he just experienced.

～

At the Chicken Chalet, a beautiful set of "bird condos" built by our carpenter Caleb Fieser, I step into one of four outdoor yards and pick up Jailbird the rooster, who relaxes in my arms during a quick lesson in chicken anatomy, physiology, intelligence, and emotions.

Both the egg industry and the broiler industry want the public to believe that chickens truly are "bird brains." As they've concentrated birds into cramped, toxic spaces hidden from public view, we humans have disassociated what's on our plate from the animals themselves. If we knew *who* they were, many of us would question what we eat; many of us would make different choices. The poultry industry is vested in our not knowing, and their success is apparent in visitors' belief that chickens are indeed "bird brains" with little in the way of personality or individual preferences. Our chickens help us challenge culturally imbedded misconceptions.

So, too, does science. Dr. Jonathan Balcombe of Humane Society University; Dr. Joy Mench of UCal./Davis; Dr. Lesley Rogers, Professor Emeritus at Australia's University of New England; and Dr. Bernard Rollin, Bioethicist at Colorado State University are among a growing number of thought leaders who have studied chicken sentience and intelligence. Science now knows, for instance, that birds have mental capacities equivalent to those of mammals, even primates. We now know that chickens recognize and remember more than one hundred other chickens, live in stable social groups, and are exceptional and determined problem solvers. And they don't just "cock-a-doodle-doo." Rather, chickens have more than thirty types of vocalizations. Roosters have two distinct calls to warn their flocks about ground predators, such as coyote and fox, and overhead predators like hawks and eagles. There's a great deal going on, in other words, in those little "bird brains."

So on our tours, some of us discuss the science, some of us tell stories from our own experiences, and some of us do a bit of both. I do a bit of both. And then I kiss the chickens, right on the beak. But we'll get to that in a minute.

"Does anyone know how to tell when a chicken is happy?" I ask, looking at the children.

"They cock-a-doodle-doo," one little girl offers.

"That's a GREAT way to tell!" I respond, since I think I agree with her. Of their many distinct vocalizations, the basic "cock-a-doodle-doo" is the rooster's equivalent of the pig "oink."

Arthur the goat has sauntered over. Arthur is the size of a small moose. Folks immediately love this beast—confident and nudgy, he wears a permanent smile. He's always optimistic that we'll dole out treats, but when we don't, it's okay. He simply stands in the middle of the group, happily accepting all the love that's offered. Plenty is offered right now: hands cover him, stroking his forehead, his ears, his back. A young girl drapes herself over back.

"Who wants to make Jailbird *really* happy?" I ask.

Several kids respond; I choose a girl of about eight. After instructions in how to safely hold a chicken, I pass Jailbird to her and ask her name.

"Okay, Samantha," I say. "Just hold him for a moment. If he gets heavy, let me know and I'll take him back."

I ask the group how many of them have cats; more than half raise their hands.

"So you know how your cat looks at you sometimes and blinks really slowly . . . almost as if she's blinking in slow motion?"

Everybody knows the languid blink that I'm talking about, and when I ask the group what they think it means, folks respond "love" or "happiness" or "contentment" or "he does it when he feels good."

"How do you know?" I ask the dad.

"Okay, I get it," he says, smiling.

"Chickens do all kinds of things when they're happy," I explain. "And one thing that makes lots of them happy, as long as they feel comfortable in your arms, is getting a neck massage." I show Samantha how to place her thumb and index finger through Jailbird's neck feathers until she finds his tiny neck, and then how to gently massage it.

The group watches as Jailbird does two "happy things"—the same slow "cat blink" and the repeated opening and shutting of his beak. We speak a bit about what humans know about chicken intelligence, and I share a few stories from the life of a remarkable rooster named Paulie, a former cockfighter who ate lunch with staff (and demanded that Alex share his sunflower seeds), went for car rides, broke up pig arguments, begged for treats, napped with my dog Murphy, spent a cold night in bed with me (sleeping on a pillow right next to my head), came running when we called him . . . and so much more. For me, what science knows pales in comparison to the remarkable experiences that CAS and other sanctuaries share anecdotally.

When it's time to move on, I take the lovely bird, hold him in front of my face, and kiss him once, twice, three times on the beak.

"I can't believe how nice he is. I can't believe it's safe to do that," a young mother comments.

"It's not only *safe* to do . . . it's important to do, I think."

"Why?" she asks.

There are many reasons why many CAS staffers are as affectionate with our animals as we are with children. First, we ask them to trust us—that's a tall order given their histories. Kissing a bird who could peck me in the eye is a way of telling him that *I trust him*. For me, it's a way of having a more complete relationship. And beyond the fact that it feels as natural to hug animal friends as it does to hug human ones, I use these special moments as teaching opportunities. Both my level of comfort with the animals and the animals' level of expressiveness often disarm our visitors. In fact, visitors' comments often begin with the words, "I had no idea . . . " The more intimate the tour, the more engaged we are with the animals, the more people make new discoveries that help erase decades of mistaken programming.

⁓

Our next stop is the main barn, where our special-needs horses and other animals live. It's the big barn that The Great Sheep Rambo watched over for eleven years; it's the one where he slept in the aisle in one of two huge piles of straw. We stop just outside the door, under the giant portrait of a steer named Samson, the words "Peace to All Who Enter Here" painted above it.

"Kids," I say. "We're about to go into a very special place, and I need your help."

We walk through examples of the term "special needs," and several children say they know someone who needs extra assistance—a grandparent in a wheelchair, a classmate, a relative who is hard of hearing. I explain that this is our special-needs barn, where animals ranging from blind horses to a timid sheep to new rescues who are just learning to trust us live so they can get a little extra attention. It's important that guests of all ages stay with the tour, be quiet and respectful, and reach out to animals only if their guide gives the thumbs up.

"Can you guys do that?" I ask.

A girl of about eight with spectacular red curls says, "I can do that, but sometimes my mom has a real talking problem." We all laugh as she looks up at her mother, a round woman with kind eyes.

"Guilty!" Mom says.

Alex is just inside the barn with a dozen or so guests. They've huddled around Hank, a huge broiler rooster who can no longer walk.

"So that's why you cannot, *cannot* eat chicken, even though ironically chicken is marketed as a healthy food," he is saying emphatically when we enter. Alex has likely walked his group through some of the key offenses of the poultry industry: the use of growth hormones that are causing human children to enter puberty early, the filth of the warehouses and the injuries and diseases acquired by the birds, the fecal count that is allowable in slaughtered chickens, and more.

When Alex moves his group forward to meet a huge draft horse named Sioux, our group moves in and takes its turn with Hank, a living, breathing example of all that is wrong with agri-business. I lift him into my arms.

"Broiler" is the term used by the poultry industry to describe their product: meat chickens. A handful of multi-billion-dollar corporations, among them Tyson Foods, Purdue, and Pilgrim's Pride, produce and kill nearly *nine billion* of these gentle birds each year. If one does the math, that's 14,000 chickens killed every minute in the United States alone. Let me clarify: that's 14,000 baby chicks killed per minute. Manipulated to grow quickly, they reach their five-pound slaughter weight at just thirty-nine days. When they go to their deaths, they still have their baby blue eyes and the "peep" of chicks.

Meanwhile, a lucky few escape their intended fate and wind up at farm sanctuaries. Catskill Animal Sanctuary has taken in hundreds. We watch as they grow larger by the day, their legs becoming three times the size of a normal chicken leg; their bodies, particularly their huge breasts, growing until the chickens are the size of turkeys. What happens then is identical to what would happen to you if you weighed five hundred pounds: They have chronic joint pain, extremely limited mobility, respiratory and circulatory issues, and more. Their organs are under constant stress. These Frankenbirds have a life expectancy of *one year*, as opposed to a naturally raised chicken, who can easily live into its teens.

Gently I place Hank in his bed of straw on the side of the aisle. At over three years old, he's an old man. Guests watch his left leg splay out, then watch as he settles into the straw.

"He can't run *or* walk?" a little girl asks.

"No, sweetie, he can't," I respond.

"That's really sad," she offers.

"Yes, it is."

We spend another hour making our way down the long aisle, out to one of our larger horse pastures, out to our sheep pasture, then finally, down the long lane where rabbits, geese, hens, turkeys, and cows live—all in their own separate barns and pastures. Along the way, guests meet Casey, the ancient paint pony who has been with us for nine years; beautiful Noah, whose story is told in the essay "If We Can Get Them Out"; The Three Wise Guys, a trio of sweet, in your face, what-kind-of-mischief-can-we-make goats; and a flock of sheep who were terrified just weeks ago but half of whom now practically knock us down when we enter the pasture. We overhear Rebecca's beautiful way of discussing the dairy industry as her tour has bunched together in front of Tucker's small herd of five cattle. ("We've gotta hold hands, guys," she says to her groups. "I find this stuff is often a shock for people to hear, because they've been told something very different," is how she begins.)

By the time the tour is done, folks have met and inter-acted with eleven species of animals. Those who wanted sat right down on the ground with me and were nibbled by sheep and nudged by goats. Others held chickens or rabbits, turkeys or ducks. Nearly everyone has kissed a horse; plenty have kissed a pig. Several have wept at the descriptions of what farm animals are forced to endure at human hands or at the valiant recoveries of horses we weren't sure would survive when they arrived at CAS. The adults have pondered the contradiction between their moral outrage toward animal hoarders and their passive accep-tance of industrialized cruelty. The kids? They have listened well. They have asked good questions. Most of all, they have given and received an awful lot of love. And at least one of them, a little boy named Dylan, has made a friend that I'll bet he won't soon forget.

Giving tours can be a challenge. As Rebecca says, "you have to be ready for *anything*, and you've got to be so many things: teacher, friend, charismatic storyteller, social worker, and activist, and you've gotta be passionate without hitting people over the head." She laughs and adds, "Each tour day kinda feels like climbing a mountain."

But tours are central to what we do; I hope that a glimpse into one of the hundreds we give each April through October has given you that greater understanding. Come join us on a Saturday or Sunday. Fair warning: be prepared to fall in love.

Animal Camp

The experiment was simple and elegant. The hypothesis—
that personality sometimes trumps species and will lead
to a better outcome for particular animals—had been
amply proven at CAS by the Underfoot Family, and more widely
proven by scores of YouTube videos documenting unlikely pair-
ings: a dog and a dolphin, a cat and a bear, a sheep and an elephant,
and so on. Still, we had never tried to introduce *big* animals of
different species. Until now, there had never been a need. We knew
that cows and horses often peacefully coexisted. Whether they'd
bond as friends, we didn't know. We had *no* idea whether a horse,
a cow, and a 700-pound pig, all ostracized by their own species,
would become friends. But we were about to find out, because
Hope, Tucker, and Franklin needed help.

Hope, a chestnut girl with a big white blaze, arrived with
four others from a failed Thoroughbred breeding operation; at best,
they were semi-feral. From the beginning, Hope stood out. In her
pasture, she would raise her head, ears forward, and look up. "Help
me trust you," is what the gesture said. Soon after the mares' arrival,

we observed a troubling dynamic—Hope got picked on. She was chased from her food, driven to the outer edges of the herd by the dominant Charmer, and blocked from entering the shelter. Whatever intervention we tried, Hope was always the outcast.

At the opposite end of the farm, Tucker, a young Guernsey steer, was struggling, too. Tucker had spent the first months of his life at a petting zoo. A visitor had fallen for him, purchasing him to prevent him being sent to "auction" (slaughter) at the end of the season. At CAS, he was placed in a paddock next to my house for a three-week quarantine period. I fell in love with the 200-pound bovine puppy, who followed me around, ate my hair, and gave me wonderful facials with his scratchy tongue.

When we knew he was healthy, we introduced Tucker into our special-needs herd of Romeo, a gentle elderly steer; Helen, a blind Hereford; and Dozer, another young goofball. But Dozer picked on Tucker relentlessly, driving him away from the herd.

"Tucker!" we called as we approached, and often he came gamboling at us, pressing his head into our bellies.

"Love me," he would say. And so we did, scratching his cheeks, rubbing his soft ears.

Meanwhile, before he was set aside to starve to death at a pork production farm, Franklin the pig had developed a defensiveness that he carried with him into young adulthood. (It's tough for the runt of the litter to overcome behaviors learned in those few weeks as the smallest of the bunch, no matter the species.) At CAS, all the other pigs picked on Franklin. Whether the issue was who got the most food, the best spot in the straw, the biggest mud hole, or the most attention, Franklin was always the loser. His desire to play also annoyed our older pigs. He shared sleeping space with Piggerty, but in their pasture, he drove her mad. *"Play with me!"* he insisted by poking her sides, doing his funny little pig dance. She whipped around and snapped at him—her way of saying: "Could I *be* any clearer here?"

So Franklin will join Hope and Tucker at "Animal Camp," (my friend David's beautiful twenty-acre property) because a) I'm crazy, and b) I sense that a friendship between a horse who lacks confidence and a needy steer and pig just . . . might . . . work?

∿

We arrive at moving day, which we know will be all about the pig. Instead of electric prods, metal pipes, and other instruments used on hog farms to load aware, terrified beings onto transport trucks, we bring a bowl filled apples, bananas, and chocolate cake, none of which will matter a whit if Franklin doesn't want to load.

At 10 a.m., our hauler backs her white Ford F-350 and glistening trailer up to the gentle slope of grass in front of the main barn. Tucker the cow and Franklin the pig will ride together in the front compartment; we'll close the divider at the halfway point, then load Hope in the back of the trailer.

Sure we will.

Like an old golden retriever, Tucker follows me on first and settles as soon as I place two flakes of hay in front of him. "What a good boy you are," I praise him as I massage his cheeks. Tucker buries his face in his hay. If only it would go this smoothly with Franklin.

Our ammunition—the food—is piled into a feed tub. If all else fails, the heavy artillery—three slices of chocolate cake—is hidden around the corner on the tire fender.

"Come on Franklin; we're going to Summer Camp!" I said, encouraging my friend to follow his nose as I back toward the trailer holding the loot. "Doesn't this smell good?" Six of us are on hand, and with bodies and voices we steer him toward the trailer . . . for about a nanosecond. Franklin wants nothing to do with the big white box. Time after time, he climbs the ramp, then retreats, climbs, and retreats again.

I sit inside the trailer, dragging the lure inches in front of an enthusiastic snout. When he reaches forward, I back deeper inside. "Ummphh! Give me the *cake!*" he demands. But I don't.

I am a fool for not thinking this through, for not having Franklin contained in a small space, for forgetting to withhold breakfast so that our food bribes will be more effective. We should have known better. When has "wishful thinking" worked with a pig? It's clear that our pink pal isn't going anywhere. The longer we try, the more agitated he becomes: full-out hysteria is about four seconds away. We agree to a "do over."

Hope loads within five minutes. With two-thirds of Summer Camp residents on board, we make our way to High Falls.

～♪

Today, we hope that Franklin, having missed breakfast, will be more inclined to step onto the trailer than he was yesterday. In fact, we are banking on that.

Corinne backs the trailer through Franklin's gate until the rear is five feet from Franklin's door. Using portable gates, we block all escape routes so that Franklin's only choices are to remain in his cushy barn, deprived of breakfast, or to follow his food dish as it backs up the ramp (because I am dragging it on the ground just inches in front of his snout) and onto the trailer.

"Kathy, we're ready to load Franklin!" Alex radios me.

I load Franklin's dish with pig pellets, cantaloupe, pear, and a pint of strawberries for good measure. Maybe these will do the trick. Holding his breakfast, I open Franklin's gate. He hurries out, following the dish that holds heaven. "We're going to Animal Camp, Franklin!" I encourage him with each step, holding the dish just in front of his face until I arrive at the trailer and place it on the floor.

I inch backward, on my knees, into the trailer. Franklin looks at me in earnest. "Ummph," he says. And whether it is hunger, fewer people to distract him, or fewer options (Franklin can only either follow me or simply stand in place) Round Two is perfect. Franklin steps tentatively onto the ramp, then follows the trail of treats until, just one minute after we've started, he is on his way to Animal Camp.

It was obvious from the get-go that David's twenty-acre paradise would be an ideal spot at which to learn a thing or two about three outcasts and what would help them lead happier lives at CAS. "Forgotten Lands" is tucked a half mile back into the woods off a rural road. One hears and sees nothing but nature from anywhere on this land. Nestled in a quarter-acre clearing near the front and graced by a dozen or so majestic oak and hickory trees, David had recently completed the most beautiful little barn I've ever seen: high ceilings, 10" x 10" posts and beams, open space, huge windows. Yes: a fine, fine spot.

"I want the office and the barn to essentially be one big room," I had explained to him as we sat on the porch sketching out our ideas, "but I should be able to close the animals into their own space when I need to."

"Why would you need to?" David asked.

"Oh, you know, a cow in an office is a fun *idea*, but things could get messy pretty quickly."

At CAS we've learned that *many* assumptions humans make about differences between "us" and "them" are based more on prejudice, ignorance, and culturally enforced separation than on truth. Animal Camp would allow me to unearth and challenge any of my own faulty assumptions. While its purpose was to build the animals' confidence, I was going to have fun in the process. I wouldn't "train" my friends to enter the office, but I *would* open the door and see if anyone *chose* to enter, and I would learn from those decisions.

Accordingly, the structure is eighteen feet deep by twenty feet wide: one big room is both barn and office. The floor is made of thick planks. An "open-air" stall occupies the front left corner and opens to a paddock; a gate can be closed to keep animals inside the stall. The other three sides of the stall are low walls, so that any visiting animal can stretch his head outward on the left into the fresh air, or the right into the office, or in the back to the feed bins just beyond his reach. It's a wonderful design with ingenious simplicity. The animals will be able to hang out on one side of the barn, I at my desk on the other, divided from them only by a four-foot wall. While I'm at the computer, I'll be able to touch a warm muzzle or a cool snout. By design, there is room for mischief and mayhem.

The quarter-acre barn paddock is connected via a gate to another of the same size. Three shelters are enclosed within this paddock. Both paddocks are shaded by towering trees, and both have gates at their far sides that open to a three-acre pasture that stretches all the way back to David's house, located a thousand feet behind the barn. If necessary, this "two paddocks and a pasture" design will enable each animal to have his or her own enclosed space.

‿

Hope and Tucker were at the gate when we arrived, all ears and attention. Unless a cow is exploding with glee, it's generally easier to read excitement in a horse or a pig than it is in a cow. But today, Tucker's insistent *moo-oooo!* told us that his curiosity was killing him. After all, from his point of view, he and Hope had arrived at David's in this same big box a few days earlier. *Who was in there now?*

"Franklin!" David and I call to our porcine pal. "Welcome to Animal Camp!" I did it to comfort Franklin. David called out to him in sheer excitement, for David has adored Franklin since Franklin's first night with us, when as a four-pound pig he fell asleep between us, snuggled into the blankets of David's big bed.

For the safety of all, David moved horse and cow into the bordering paddock. I swung open the trailer's heavy metal gate. "Hello, my love," I whispered to my friend, who turned around from his forward-facing position and ambled toward me. "Come look around!" He was shaking.

At CAS we pulled homemade ramps up to the trailer to ease Franklin's loading; until now I hadn't even thought about how hard it would be for my short-legged pal to take a huge step down to solid ground.

"Oh, wow," I said, looking at the big step, then at Franklin's little legs. "This is going to be hard for him." I was wrong. Franklin simply turned sideways, his body parallel to the opening, and lay upright as closely as he could to the edge of the trailer, his front legs outstretched. Then, with the bulk of his weight resting on the trailer floor, he scooted forward with great effort and placed his left foreleg on the ground. He scooted again and placed his right foreleg.

"You're a smart boy, Franklin," I praised him. Franklin rotated his body until his butt sat on the edge of the trailer, then stepped down with one rear leg, then two. He walked several steps, then lifted his pink snout to the air to take in his new world.

"Look Hope, look Tucker," I encouraged the onlookers. "This is your new friend." Hope was a nanosecond from fleeing for her life, but stood in place, her eyes growing larger by the second.

"Moo-oooo!" Tucker called.

Franklin, though, was more interested in his home than his housemates. After all, this pig had been picked on his whole life. Today, his snout reaching no higher than the elbows of the red horse who towered above him on the other side of the fence, Franklin was keeping his distance.

A whole new world was laid out before him. Water and ferns filled a low spot thirty feet from the barn entrance, and the moment he saw it, Franklin became a pink tank moving toward it. He rushed into the water and shoved his snout to the ground. Up came a face covered in cool brown mud. Again, this time deeper. Franklin's snout was a shovel, and each time he lifted his face, he loosened more heavy wet earth until the space was precisely the way he wanted it. He dropped his body down into brown, goopy bliss.

"Pig dipped in chocolate." David laughed when a few moments later, Franklin stood to continue exploring.

"Ummmph," he snorted a greeting as he walked past us.

"Hey, happy pig!" I responded.

For the next few minutes, Franklin explored every inch of his paddock. He smelled the pine trees, rooted at the edge of the fence, blew bubbles in his water trough, and discovered a new delicacy: hickory nuts. For a good ten minutes he searched for them, snout to the ground. When he found one, he'd gobble it up, crack the shell and spit it out before chomping on the tasty nut, *ummpphhing* his approval with every few steps.

"What do you think, Franklin?" we asked him.

"Ummmpph."

Soon, his misgivings notwithstanding, Franklin was curious to meet the beings locked behind the gate. I stood up from my perch and walked over to encourage the introduction. Red horse head reached over the fence as pink tank approached. Hope's neck was arched, her ears were forward. She was anxious but eager, the desire to say hello transcending species. And then there was Pink Tank. My dear Franklin, who had to overcome three years of being jostled and jolted by porcine pasture mates. Who had to overcome being bitten by Maxx the horse when he strolled through the barn. Did Franklin have the capacity to simply let the moment happen? I am delighted to report that yes, indeed, he did. Muddy pink snout lifted, soft black muzzle lowered, and in a moment so pure that it is forever etched in my memory, two creatures abused by their own said hello. They simply breathed, taking each other in. Nothing more than that; but for these two, it was an auspicious beginning.

⁓

I am filling the water trough when our friends Mary and Lola pull in. It's been two days since Franklin's arrival. Lola, a scrumptious three-year-old, has been to CAS often. Now, however, Franklin is her neighbor; Lola and her parents live just a couple miles up Mohonk Road.

"Lola!" I shout as I walk toward their car. Lola buries her face in Mary's neck. "Would you like to come see *Fanklin*?" I ask, pronouncing his name without the *r*, the way Lola does.

Not a peep from the child.

As Mary and I chat, though, Lola signals to Mama that she wants to see *all* the animals. The three of us walk into the open air barn, where Hope is resting in the only stall, separated from us by a four-foot fence. I climb over, inviting Lola to come, but Hope's mass is too much for her. She stays in Mary's arms. The good horse presses into the fence, stretches out her long neck and gently touches Lola's calf. Lola stiffens.

"It's okay, Lola," Mary encourages her. "She's saying hello."

Slowly, Hope reaches up and nuzzles Lola's cheek, signaling that she's really just love in a thousand-pound package. It works. In a moment, Lola is in the stall, brush in hand, grooming the sleek red body of the animal who towers above her. Tucker, known by Lola as "Tucka," is hanging out just outside the barn, his legs curled under him as he chews his cud. (Despite a field filled with grass, Tucker always chooses to be near me. If I'm in the barn, it's where he wants to be. His preference, in fact, is to be *in the office*, right next to me, but his office skills leave a little to be desired.)

"Hi, Tucka," Lola says to her red and white friend, and we walk over with some bananas that we've grabbed from the fridge. Meanwhile, Franklin is taking in the action from a nearby shelter. As we move toward the trough to finish filling it, he barrels over, steps gingerly into the trough, and sits down, sloshing most of the water out and instantly muddying what remains.

"Franklin! This is not your bathtub!" I say to the imp, who is clearly showing off. In fact, I'm pretty sure he's doing this to get a reaction from Lola. When Lola laughs, I turn to her. Forcing earnestness, I say, "Lola, this is not good. This is water for Franklin and his friends to drink. What do you think we should do?"

When Lola is excited, her voice could wake the dead. Her suggestion is instantaneous and resolute: "We should say, '*DON'T GET IN THE BATHTUB, FANKLIN!*'"

With that, Franklin steps out, turns around, roots under the trough, and very deliberately tips it over. We laugh together as I walk them back to their car. Lola is climbing into the Subaru's back seat when Mary says, "Lola, want to tell Kathy your big news?"

"*I DON'T WANT TO EAT MEAT ANY MO-WA!*" Lola hollers from the back.

"Lola, that's fantastic!" I praised her. "What made you decide to do that?!"

"BECAUSE I LOVE ANIMALS!!" she hollered.

"I DO, TOO!" I yell back.

And then they are gone.

⁓

It's seven o'clock when I walk through the barn and into the pasture. No hint of anyone. Normally, the threesome is hovering around the office when I arrive. Occasionally, though, I have to call them in.

"Hope! Tucker! Franklin! It's breakfast time!" I holler as loudly as I'm able.

Hope is always first, of course: she's a *Thoroughbred* after all. As soon as she enters the field, she breaks into a gallop. She's covered three hundred feet and planted herself in the open stall when I spot the pig and cow pair trotting through the lane. For a moment they're side-by-side. Tucker is all goofy-cow now. It's not that cows necessarily look funny when they're moving quickly—it's that cows moving quickly when they're excited are, well, *hilarious*. They don't merely run. They run and buck and jump and hop and toss their heads and kick their rear leg sideways and try their damndest to do it all simultaneously.

So anyway, here comes Tucker, hauling ass. Beside him, at least for a hundred feet before Tucker can outrun him, is Franklin, running as fast as his stump legs will allow. I stand there, urging them on. It's a delightful way to start the day.

I turn to the fridge and pull out the bucket marked "Franklin's Food," and finish slicing the morning's offerings: watermelon, summer squash, chard, strawberries. The troops register their impatience. Hope leans over the rail, stretching her neck and head as far as they'll go, flapping her muzzle for a treat. She sees the produce that I'm chopping for Franklin, and she wants, well, all of it. I've never known a horse as eager to try new tastes. Hope is a true vegetarian gourmand. I slip her half a peach. Just behind her, Franklin is *ummphing*, and the longer I take, the louder he becomes. Tucker, meanwhile, is doing what cows do as well as dogs: He's staring at me. His head, too, hangs into the room. He's hoping for a treat, for in his six weeks at Summer Camp, he's discovered kale, plums, mangoes, bananas, strawberries, and whole wheat organic tortillas, and has perfected the art of respectful begging. He knows that all he needs to do is stand at the gate, six feet from the magic white box out of which emerges the food of the gods, and stare at me with all the earnestness a cow can muster until I am no longer able to resist. We both know the game.

"Okay, Tucker, yours is coming," I say, and I take him a bunch of Swiss chard, offering it one huge stem at a time. He is delirious.

I place Hope's dish in the stall, and Franklin's squeal rises to a deafening pitch. He knows the routine: Hope first, Tucker second, Franklin third. The squeal is how he manages the wait. If he couldn't squeal, he'd probably need to plow down a tree. I reach over the low office wall and carefully toss Tucker's dish to the ground. He dives in.

Franklin follows breakfast to the paddock that houses the run-ins. Here, he can relax as he eats. About a week ago, Tucker figured out that Franklin is a 700-pound pushover, and as soon as he finished breakfast, he charged over to Franklin's dish and dove in. Franklin stood his ground for a moment, but eventually retreated. So now Franklin eats alone—and licks his bowl clean.

I grab a pitchfork, shovel, and rake, and while my friends are feasting, I pick up poop deposited the night before in the largest of the shelters and toss it into a pile that David will retrieve with the front-end loader at the end of the day. There is horse manure, cow manure, and, in the front corner of the shelter, pig manure. I smile. David designed camp so that each of the animals could have its own sleeping space, but last night, during a thunderstorm, they chose to be together.

⁓

Both immediately before and immediately after breakfast and dinner, one of two things happens: either the three friends graze together in the shared pasture, or, if I'm working in the office, they hang around, wanting to be as close to me as possible. Frequently, two of them will share the stall just next to where I'm working, and the third one, usually Tucker, will stand just outside the office gate, hoping to be let in. I'd let all of them in far more often than I do if there weren't an obvious downside to having a horse, a cow, and a pig as office assistants. I've done it, several times. Hope and Tucker mess with everything: the grooming box, my stack of papers, the halters and lead ropes, the phone, my laptop; Franklin, though he's generally well-behaved, is still a pig, and could easily flip the table that serves as a desk in an effort to get to the feed bins just beyond his reach. When the animals are in the office, let's face it: I don't get too much work done. Mostly, they have fun while I put the computer out of reach, encourage their exploration, and acknowledge that it's a good thing that I don't have children. I'm sure they'd all be criminals.

It's a muggy day, and Franklin heads for his mud hole. I walk over with the hose to fill it with more cool water. First, though, I spray his pink body, and Franklin grunts his pleasure.

"You're welcome, pig!" I say to him and watch as he lowers himself into the cool earth.

Both Hope and Tucker are clammy—a perfect excuse for bath time. Six weeks ago, the hose frightened Hope and Tucker would hightail it away from me as soon as I turned the water on.

Now, bath time is a big event. I gather shampoo and grooming tools and place a blue halter over Hope's red head. I pick up the hose, adjust the pressure, and begin at her hooves. Quickly, I work my way up the sleek body: leg, shoulder, neck, back, rump. I spray with the left hand, massage with the right. I don't need to hold her for her to be still. Hope loves this. I move to her head; this has become her favorite part of our ritual.

"Hey, good girl," I encourage her, adjusting the pressure again to a fine spray as I mist her forehead. She closes her eyes and stands motionless, enjoying the experience, so I adjust the pressure upward and massage her ears, her forehead, and her eyes as the water runs down her face.

A few minutes later, Hope has been scrubbed, soaped, and rinsed. I praise her, remove her halter, offer a banana, and watch the inevitable as she trots to the pasture, rolls on the ground, and is instantly filthy. Tucker, standing nearby the entire time, allows me to place the same blue halter over his white and red head. This act is a victory in and of itself, but it's nothing like giving a cow a bath and watching him love every moment, particularly when just three weeks ago, he ran for his life when I picked up the hose. I spray a hoof as he chews on my shirt sleeve, and as I move the spray up his legs to his neck and shoulders, he stretches my shirt until it could fit a linebacker. "Stop, Tucker!" I say, but I'm laughing the entire time, so it's not like he's actually paying attention.

Later that night, David and I are sitting on the porch, watching the sunset. My old dog (and best pal) Murphy is in the grass below us, sniffing the air for deer. He knows they settle all around us for the night: on the cliff to the left of the house, in the berry thicket beyond the two ponds. A sound to our right is far too loud to come from a startled deer. We look over at the pasture, and here come the three musketeers: Hope and Tucker are running; behind them by fifty feet, Franklin is doing his best to keep up. They come to the gate just fifty feet from David's porch, form a line, and look over at us. David retrieves carrots and bananas from a bowl in the kitchen, and the three of us walk down to greet the three of them. We are unlikely friends, but we are friends nonetheless.

The purpose of Animal Camp was to see how three "victims" would coexist and how that experience would change them. It was to figure out whether it is sometimes preferable to group animals by *personality* rather than by species and to take back to CAS lessons that might impact how our animals live. And so, at summer's end, we close up the barn and return to CAS. I am delighted to share what I have learned.

Within a couple days of Franklin's arrival, Hope took on the alpha role, and while she used her newly gained status at feeding time, she also used it anytime the gangly, rollicking Tucker approached Franklin a little too assertively. "Play with me!" Tucker insisted as he charged at Franklin, head down, legs whirling through space. Did Hope sense Franklin's panic? Did she empathize with him? All I know is what I witnessed. Time after time, Hope placed her body sideways in front of Franklin, physically preventing Tucker from intimidating him. Tucker quickly got the message. It seems the maternal role suits Hope well.

Hope will return to CAS a more confident and trusting horse. Her uncertainty is gone, as is her dislike of having a halter placed on her head. I can pick up her hooves and lean into her as she naps. I bathe and groom her and watch her body relax rather than stiffen. Hope has also benefited from being the undisputed leader of her tiny pack and sharing her days with two pals who enjoyed her company without an agenda. I wonder if she will carry her newfound confidence with her.

Had the sensitive horse not been here to figuratively hold Franklin's hand, I'm not sure Franklin could have broken through his fear of Tucker. Prior to this summer, Franklin never had a protector. But now, he knows in his bones that neither Tucker nor Hope will hurt him. How do I know this? I know this when I see the three of them grazing in a tight cluster. I know it when Tucker comes barreling right at his porcine pal and Franklin stands his ground. I know it when the three of them are jammed into the tiny run-in David designed for a single animal, Franklin lying in a mound of hay with Hope and Tucker hovering over him. For the beleaguered Franklin, this level of trust and confidence approaches miraculous. It remains to be seen whether it will serve him in a group of pigs.

And Tucker? Hope has taught him the concept of limits. Hope has taught him to slow down and take a breath. Tucker understands that he can't play with Franklin in the way that young steers like to play. He's tried. These days, he approaches Franklin slowly, stretching out his neck until cow nose and pig snout meet. They stand that way for a few seconds, saying hello, Tucker saying, "I'm your friend, pig. You're safe with me." Tucker has also benefited from the unique nature of his summer pack. At Animal Camp, there's no vying for a place in the group like there is at CAS. The group just *is*—an affable, uncomplicated unit. Finally, this always affectionate young steer now has the personality of a golden retriever. Happy for attention since the day we took him in, he's come to expect it after daily grooming, frequent bathing, trips to the office, and nighttime snacks.

Tucker and Hope nap near Franklin as he squishes into his mud hole. They spend a good deal of their days napping, in fact, and while they sometimes do retreat to their individual shelters, they just as frequently cram into the one designed for a single pig. All day long, whether grazing, resting, or simply "being," these three keep each other company, respect each other's limits, and prefer closeness to separation. In terms of the animals' happiness, Animal Camp has been a resounding success. So, too, regarding the larger purpose of the experience. If one can generalize from a single experiment—if Tucker, Hope, and Franklin have proven that in some cases, personality, and not species, should determine who lives with whom—then we've got some thinking, and perhaps some animal rearranging, to do.

Returning Home

Hope and Tucker have traveled quietly back to Catskill Animal Sanctuary. We're placing them in the large pasture behind the barn—the field where the special-needs horses spend their days grazing peacefully together. At dinnertime, we retrieve them—Noah, Star, Freedom, Abbie, Lexie—and they return to their individual stalls for the night. It's a kind, benign group. Will this group of special-needs senior horses accept our newly-confident Hope? And if this happens, will she help Tucker integrate with the horses? This, after all, was the main purpose of Animal Camp: for three outcasts to bond and blossom so that they could return to CAS and live more happily here. Other than alone, we've nowhere else for Tucker to be—he'll be ostracized by the larger or more aggressive cattle.

I hand Hope's lead rope to Allen. I take the gangly cow, who leaps the ditch and half drags me to the gate. Thankfully, the other horses are at the far end of the field, oblivious to the goings-on. "You're a beautiful girl," Allen says before releasing the mare; I hug my steer before I let go. "I love you, Tucker," I say to him.

Hope eyes the horses below her. She arches her neck and trots toward them. At once, they move forward to meet her. First, Noah. The others follow closely behind, and in an instant we're watching an equine dance improvisation: the nose-to-nose greetings, the circling and snorting, the charging away and coming back. I am struck that there's not a single kick, not a single squeal; instead, just the raw beauty of a group of kind horses connecting with the newcomer.

Tucker, watching from a distance, is beside himself with excitement, and when the herd takes flight as a single group, Tucker barrels forward to join them, head low and back legs kicking out in unbridled cow joy.

But then the worst possible thing happens: Hope turns on him. She charges menacingly, ears back and teeth bared. She means business and Tucker knows it. He backs off.

The horses settle within a few minutes, seven heads down to graze. Tucker again draws near, but then three horses, one of whom is Hope, come at him, and he is suddenly running for his life, right toward the humans. Yes, Tucker is racing toward us with three horses behind him. Humans rush forward with menacing voices and hands in the air to redirect the horses. "Tucker, I'm so sorry!" I call as the young steer careens forward, stops dead, then attempts to hide his 800-pound body behind my 120-pound one. Satisfied, the horses saunter off, leaving the frightened, dejected young steer, who lowers his head, pressing it into my calves, asking for a good rub-rub-rubbing of his cheeks, his forehead, his ears . . . the kind he got every day at Animal Camp.

When I enter the field a half-hour later to check on my friends, I have a curry comb in my left hand, a brush in my right. Hope is in the middle of the herd, her head deep in late summer grass. Tucker is grazing by the pond, three hundred feet away from the horses. I call to him. He looks up and walks toward me, uttering a low, baleful *moo-oooo*.

"My little man," I say to him as he lowers his head for another rub.

"Moo," he responds. "Moo-oo!" Tucker is talking a blue streak right now. "Mom, we were best friends all summer! We slept in the barn together and grazed together and at night we walked up to the house together, and in the morning when you called us for breakfast, we galloped through the field and made you laugh every time! I thought she loved me, Mama . . . I thought she loved me." I feel like my young friend is saying all this and more as he continues to wail. "Mmmmm," he moans as he lowers his head, pressing into my thighs.

I move the round, rubber curry comb in circles over Tucker's cheeks.

"Moo," he tells me.

"I know, Tucker. I didn't expect this either. I'm so sorry," I say.

"Mmmm," he continues.

"We'll figure something out," I promise, before heading to High Falls for Franklin's return trip.

~~~

Ninety minutes later, Franklin steps out of the trailer and walks into our spacious new pig pasture. With a Plan B available, we've decided to give the previously uneasy relationship between Franklin and his former roommate, Piggerty, another chance. Piggerty despises big pigs Jangles and Farfi and is ignored by new arrivals Roscoe and Reggie. "She's been depressed all summer,"

April explains. We hope that Franklin's newfound confidence will serve him. If he stands up to Piggerty, there could be a happy truce. If not, we'll move to Plan B.

Franklin's head is low as he explores his old turf, checking out the boulder pile and water trough, rooting in the dirt. Piggerty is at the far end of the field; she has not seen him. Franklin crosses the creek, steps up the bank, and heads toward the new pig barn. I follow closely behind him. He steps over the threshold and sniffs the air. He looks at me, acknowledging the new space that smells of just-hewn wood and fresh timothy hay.

While Franklin explores, I step back out into the field and see Piggerty making her way toward the barn. Allen is rounding the corner on the tractor, and as he pulls it over to park it under the willows, I call to him. "We might need you. Can you come in?"

Allen enters the field just as Franklin exits the barn; Piggerty is only thirty or so feet away. They have seen each other. There is no need for words here; Allen and I know that this could go one of several ways, and we know what to do if the worst happens. (Animal introductions differ tremendously by species, with rabbit and pig introductions being the most uncertain, complicated . . . even dangerous.)

Franklin and Piggerty are now snout to snout. Franklin talks. Piggerty is silent—not a good sign—but the playground bully seems happy to see her former mark. They dance around each other, then turn away. They circle together, Franklin leaning into Piggerty's bulk. They walk side-by-side, each eyeing the other, and just as Allen and I say to each other, "I think it's going to work!" Piggerty pounces, digs into Franklin's flesh, and Franklin is running for his life.

Frantically, he darts behind the rock pile with Piggerty on his rear. Allen and I split up—I to cover Franklin, he to discourage Piggerty—but at the soonest possible instant I open the gate wide for Franklin's escape. Most pigs need to be given room and time to work out their issues (we prefer they do this under supervision), but Franklin is *always* the victim. With so many more living options for him than there were prior to his departure, I'm not prepared to risk his newly acquired confidence.

"Come on, pig," I say to my pal as I open the gate. "It's time for Plan B."

We introduce Franklin to the potbelly field across the road. Shy Girl, Mabel, Charlie, and others live in this field with its spacious heated barn. I sit and watch. Franklin is uninterested in the potbellies; they are equally uninterested in him. The dominant Muffin makes no effort to confront Franklin, while Franklin simply settles in immediately, going right to work exploring his new digs. Fascinating—apparently, among the porcine set, size matters a great deal. Without exception, when we've introduced pigs to other pigs of similar size, the introductions have been epic. For Franklin, the little pigs may as well be pottery.

These, then, are the pigs with whom Franklin will sleep. The next piece of the plan involves Franklin sharing the adjoining field, where Hope has successfully integrated into the herd of horses. Tucker is now grazing on the edge of the herd; I suspect that soon, he'll have worked his way in. But if the group is kind enough to accept Franklin, or at least to ignore him, then all three Animal Camp friends win: Hope can hold her own; Tucker, who has never shown much interest in other cows, will be right behind the barn, where it will be convenient to lavish him with attention; Franklin will also be nearby, with a huge field to explore, human friends with whom to interact, and maybe, just maybe, his bovine buddy to pal around with.

We open the gate wide for Franklin to enter the horse pasture. The horses are clustered some two hundred feet away. Hope is in the middle; Noah and an old mare named Crystal are closest to us. Tucker is only thirty feet or so from the herd, resting under a small tree at the edge of the pond. I walk halfway toward the horses, fold my legs, and sit down to watch what unfolds; I'll stay until it's clear that I'm not needed.

The next half-hour passes uneventfully. Franklin snuffles and roots, walks a few feet, tears off a mouthful of grass, and snuffles some more. The horses' heads lift as one by one they spot him, the pink tank proceeding through the pasture, driven by his nose. But that's all they do: they see him, they look for a moment, then put their heads down to graze. *Whew.* When Hope spots him, she looks for a moment longer than the others, but doesn't leave the herd to say hello to her summertime pal. Only Noah, the lone gelding, looks for a good long while, then turns his head around to look at me. "Is he supposed to be in here?" I think he's asking.

"Thank you, Noah!" I call to him, then move toward him to acknowledge his gesture with a good long head scratch. "That's Franklin. He's a good pig." Other horses would likely terrorize pigs. But this is an uncommon group of generous spirits, and the introduction proceeds without incident.

Just a few yards from Tucker, Franklin either sees or smells his buddy, and in an instant is barreling toward him. "Mmmm-phh!" he calls.

I stand there watching as Tucker licks him, remembering, and I, in heaven, remember the first time white cow muzzle met pink pig snout.

It has been a good and worthwhile summer.

Moments of Glory

In 2004, Aries arrived at Catskill Animal Sanctuary with forty other animals from a failed sanctuary. He was a small, beautiful sheep, with loose, curly wool, enormous, penetrating eyes the color of wheat, and a gentle, unassuming nature.

For years, Aries lived with best pal Lumpy and the rest of his flock in a large pasture at the northern end of our property. In ones and twos, the flock shrank: a pair was adopted, then another; two elderly sheep died. And throughout those same years, both Lumpy and Aries became old men who eventually earned a coveted spot among the Underfoot Family, the ever-changing group of animals who live in our main barn and who, during the work day, do essentially whatever they choose. It always fascinates me that with a large farm to roam, most of them, no matter their species, choose to be underfoot.

Among all the creatures who reside at Catskill Animal Sanctuary, we know the Underfoots most intimately. We weave through them—these sheep, goats, chickens, turkeys, pigs, even the occasional horse—as we clean and bed stalls, rake the barn aisle, lead other animals out for the day, load vehicles with hay. We shoo them from the kitchen after they've maneuvered their way in, hoping to raid feed bins; we kneel to offer kisses or strokes of the head to goats or turkeys peacefully resting in a pile of hay.

For the last two years, Aries has been a beloved member of this family. In a gang of enormous personalities (the boisterous and verbal turkeys Ethel and Henrietta, drama queen Hannah the sheep, crooked-legged Atlas, the special needs goat, and others), the kind and self-possessed Aries was a refreshing counterpoint: tranquil, self-possessed, a perpetual smile on his face.

～つ

It had been two days, both bitterly cold, since Aries left his stall; two days since he ate any food offered him. This morning, though, was warm and sunny, and staff member Sara marveled as she watched Aries struggle up on shaky legs, hobble down the long barn aisle, and slowly circle the barn.

"Kathy, I'm not at all one of *those* animal people," Sara said to me later. "But it was pretty clear that he was saying goodbye."

Along his way, he must have spoken with his friends, for when he returned to his stall to rest, they followed him in: Barbie the hen, Henrietta the turkey, Atlas the goat, and Lumpy, who had barely left his side in the two previous days. Barbie the hen lay no more than three inches in front of Aries' face. We humans have no idea how they knew, but it appeared to all of us, not just Sara, that the entire Underfoot Family did indeed seem to know that their friend was leaving.

⁓

When I walked in the barn, Dr. Rosenberg was standing outside Aries's stall. Aries looked quite comfy in a pile of fluffy hay; staffers Abbie, Michelle, and Sara, and volunteers Teddy and Andrew were with him. When I walked to the door, Aries stood and moved purposefully toward me until we simply couldn't be any closer, and I was overcome by the love that radiated from the gentle beast. That he was aware of and at peace with what was imminent was not unusual—others before him have been as well. What was unusual were his efforts, all morning long, to say goodbye to the farm that was his home, to the beings who were his friends.

Michelle confirmed this. "He's been doing this all morning," she explained. Making the rounds. Acknowledging relationships. Saying goodbye.

"I love you, friend," I whispered.

As Aries grew sleepy from the initial injection, some of us offered quiet memories, words of love. Others simply held him. When he was ready, Mark gently shaved a small patch of wool from his neck and injected the solution that would stop Aries' heart. More soft words and tender hands enveloped the old sheep.

Though it had been many months since Aries could walk without pain, on this final day, it took five times the normal dosage for a sheep his size to still his mighty heart.

⁓

For a few hours, we left Aries's body in the stall so that his animal friends could do what they needed to. When we returned from lunch, Atlas, the special needs goat, was lying between Aries's front and rear legs, his head resting on Aries's belly. The birds rested in a semi-circle around the two friends.

On his final day, an unassuming sheep with a perpetual smile showed us what he was made of. Peace. Wisdom. Empathy. Dignity. Gratitude. Astounding grace. When my time comes, I hope I possess the strength of character to leave as he did: acknowledging and thanking those I loved for sharing my journey and letting them know that on the next leg, I will be just fine.

A Girl Named Norman

Like millions of turkeys every year, Norman was destined for the Thanksgiving table. But thanks to an interesting twist of fate, some well-timed phone calls, and a few soft hearts, Norman celebrated Thanksgiving with the rest of the crew at Catskill Animal Sanctuary.

When WSPK, a Beacon, NY-based radio station, advertised "turkey bowling" in its parking lot, the calls and emails, all of the "You've *got* to save the turkeys!" variety, poured into our office. *Surely* the event was a prank. Curious and concerned, though, business manager Julie Barone and I drove down to Beacon, video camera tucked beneath Julie's arm. (Many, many thanks to our "Anything for the animals!" Julie, who woke up way, way too early for her liking.) There, though the on-air DJ bragged about "the crowd," a mere seven spectators stood in the cold, waiting to bowl frozen turkeys down an alley of plastic garbage bags at ten pins borrowed from a nearby bowling alley.

A blue SUV pulled up with "Norman," the frightened turkey borrowed from a nearby turkey grower, in the back. As if the entire event weren't nonsense enough, the station evidently also needed a live prop.

"Who's gonna get him out?" a heavy woman with heavier makeup asked.

"It's a *she*, not a *he*, and you need to be careful. She's already terrified," I said. Julie stood just behind us, recording the spectacle. A guy reached in, grabbing Norman by the wing.

"That's not how to hold a bird," I stated flatly. "Wrap your arms around her so you can pin her wings and support her weight. Otherwise, you'll both get hurt."

Norman and her cage were set up between two loudspeakers. The DJ spun the story, describing how the turkey was having fun, the crowd was having fun . . . golly gee weren't we all having *fun?* Meanwhile, all Julie and I saw was a terrified bird, seven cold people, and three Butterball carcasses waiting to slide down a "bowling alley" composed of plastic garbage bags. We knew the torturous life and death of the Butterball bird; undercover videos by Mercy For Animals at various Butterball plants have exposed barbaric treatment. We hoped to prevent at least the "death" part for Norman.

﹌

So there we stood in the parking lot—twelve people if you included Julie, me, and the radio staff—and a frightened, hyperventilating turkey locked in a crate.

"I'd like to take this bird to Catskill Animal Sanctuary," I said to Jason, the station manager, who had been called outside because two strange women were more interested in the *live* turkey than they were in the competition.

"He's not ours," Jason explained. "He's the property of Quattro's Poultry Farm. And what's Catskill Animal Sanctuary?"

I explained that CAS was a haven for abused farm animals and that this was an abusive situation. "And the turkey is a *she*, by the way."

"Look," he said, his eyes dropping. "I just wanted people to have fun. It's a holiday. It's supposed to be festive."

I softened. "Does it look like they're having fun? You've got only seven people here, and three of them look like unless they *win* the competition, they won't be having Thanksgiving dinner." I motioned to a chain-smoking mother and her two gaunt young daughters, all underdressed on this frigid day.

Jason hesitated a second before giving me Quattro's phone number. "What they want to do with the turkey is their business."

"Thanks," I smiled. "You know, you might rethink this event for next year. You're welcome to come celebrate with us—people *will* have fun . . . and so will the animals."

"Sounds good," he said.

∿

"Sure, you can have the bird," said Carmen, evidently the owner of Quattro's, over the phone. "You'll have to buy it."

For lots of reasons, I'm not a fan of purchasing animals to save them from being killed. While we're contacted all the time to save animals on their way to auction, for instance, we usually decline to intervene. Why put money into the hands of someone who will simply buy more animals to grow for slaughter? That's how the thinking goes: through one's purchase, one continues the cycle of suffering and injustice.

But this was an exceptional situation. "Norman" had some degree of notoriety, as the radio station had been hyping its "turkey bowl" for weeks. If she could bring guests to CAS to discover that turkeys, cows, pigs, chickens, and other animals that most humans eat are remarkable in their own right, then we needed to find a loophole in our "no purchase" policy. Julie and I pointed the car in the direction of Pleasant Valley.

A huge sign on the porch of Quattro's old clapboard general store read "Fresh Killed Chickens." I stepped inside. A line of people waited at the single cash register. Each person held a newly slaughtered turkey. Some had geese, ducks, and pheasants as well. At the back of the store, guns, ammunition, and camouflage gear lined the shelves.

"Hi," I said to the cashier. "I'm looking for Carmen."

"She's at the counter," she said, pointing behind her.

Another long line. It was, after all, the day before Thanksgiving, and this was *the* place, apparently, if you wanted "fresh killed birds."

A man easily weighing five hundred pounds hoisted each package to its eager recipient, who then proceeded to the cash register.

I approached him. "Is Carmen here?"

An elderly woman walked toward me. "Kathy?"

"Yes. Hi, Carmen."

She was a small, bent woman, around eighty years old. Though her hands were gnarled with arthritis, they were strong hands. Carmen was a worker. She came toward me and took my hand, pulling me to a screen door. We walked into a pantry, away from the eyes and ears of her employees.

Carmen looked up at me. "I love animals," she whispered. "I love all animals. I love these birds. I wouldn't do this if I didn't have to."

I could have said *so* much in that moment, but instead said only, "Why don't you come visit Catskill Animal Sanctuary?"

"Yes. I'd like to do that."

I went to the car to retrieve a brochure, and on it wrote my name and phone number. Carmen returned to her place behind the counter. I walked out, hurting not just for the millions of birds senselessly slaughtered for this one holiday, but also, somehow, for the person responsible for many of those deaths.

At the bottom of the drive, a stressed-out Norman paced in the same cage that had held her between the radio station's speakers.

"Hi. We're here to pick her up," I explained to a toothless gentleman who approached our car.

"I'll get her for ya," he offered, and before I could span the few steps between the car and the turkey, he reached in to drag her out by the feet.

"Please, let me do it," I said, pushing myself between him and Norman, her terror rising again. I squeezed my upper body into the crate, hovered over her, pressing my chest into her back and wrapping my arms around her sides.

"It's okay, girl," I whispered to her as I cupped her little head to get her safely from the crate.

"You're safe now. Time to go home."

Julie and I settled Norman into the rear of my Subaru wagon in a large crate thick with straw, then began the slow drive back to her new home. There, on the day for which forty million birds are slaughtered while those responsible give thanks, one single turkey will discover what it's like to have room to explore, a spacious barn in which to settle down each night, friends if she chooses, and the freedom to determine how she spends her days. One single turkey will be allowed to have her life.

∿

One of the most striking discoveries in twelve years of doing this work is the individuality within each species. I'm sure I knew this intuitively, but to experience it is something altogether different. It's yet another way that farm animals are similar to, not different from, their exploiters. At Catskill Animal Sanctuary, this issue matters to us. The more each animal evolves, the more effective she will be as an ambassador for her species. The suffering of our food animals doesn't easily reach the person who doesn't care about cows or pigs, turkeys or chickens. At CAS and other farm sanctuaries, the animals do their part to open hearts, which is often the first step in changing entrenched, reflexive dietary habits.

The prevailing view, of course, holds that our species is entitled to do what it pleases *to or with all other species*: might makes right. In the case of food animals, we're simply entitled to grow them for food—and to torture them throughout that process in the interest of profitability. We're entitled to wear their skin. We're entitled to profit from all their inedible or unwearable parts by grinding them up, boiling them, adding chemical and color and scent and turning them into food for other animals, personal care products for humans, and more. The notion that perhaps farmed animals exist for their own reasons and are entitled to happiness certainly goes against the cultural grain. Yet many, most notably Dr. Will Tuttle in his important work *The World Peace Diet,* argue that this unthinking, wholesale, systemic violence is the root of all violence in the world, and that until we as a species recognize that fact, the world will never know peace.

Let's consider, for a moment, the forty million turkeys just like Norman who were not lucky enough to find sanctuary. What had their lives been like before they wound up on your celebratory table? From the "milking" (masturbation by humans) of the males, to the shackled insemination of the females, to the tearing and slicing off of various body parts without anesthesia, to the rate at which young bodies are forced to grow, to the fact that as many bodies as can fit are grown in sterile warehouses where the filth causes illness and infections and the ammonia in the air causes eye and respiratory distress, to the abomination of transport—animals yanked up by the first body part that handlers can grab, animals thrown by the wing, legs, or neck into crates, jammed in as tightly as if they were Post-it notes—to the number who die from temperature extremes, to each horrible step at the slaughterhouse . . . Perhaps these birds feel nothing but relief as life bleeds out of them in the slaughterhouse "bleeding tunnel."

For sure, Butterball paints a very different picture for us each Thanksgiving. Please, if you do nothing else, at least acknowledge the truth. Their lives are wretched. If you've seen Mercy For Animals' under-cover footage at Butterball plants, you got a peek into those lives—torture and terror from beginning to end. Through it all, they are babies. We don't identify them as such, but turkeys live just fourteen to eighteen weeks of misery. That's all. Just like veal calves, just like the billions of chicks slaughtered at seven weeks old, we kill forty million baby turkeys every Thanksgiving. I wonder: Does this sit okay with you?

Today, we've invited Norma Jean (as Norman has been renamed) to join the Underfoot Family, as she is, for the moment, our lone turkey. I'm having just a little too much fun watching her discover this big new world.

It is always a privilege to watch each new animal make the choices that so many others before her have made: how far to venture out from the safety and certainty of the barn, whether to seek out human friendship or keep her distance, which animals to befriend. Will Norman Jean, I wonder, become a larger than life, iconic figure like so many of our birds? (Henrietta, for example, whom we called our "wild child," talked and talked and talked . . . to anyone who would listen. She was devoted to Atlas, a loving special-needs goat, and often left his bed only to chase amorous roosters out of the barn when they were harassing Henrietta's hen friends. Former animal care director Abbie Rogers called her "our token feminist.")

"Come on, girl," I say quietly to Norma Jean as I open her stall door wide. "Come meet all these new friends!"

Norma Jean's only world to date has been filled with birds like her. Turkeys and indifferent (or worse) humans, more than likely, are all she has known. No surprise then that it takes her more than ten minutes to gather her courage, but finally, here she is, standing in the threshold. Her round eyes blink. She is motionless except for her head, which moves in every direction to take in the big world before her. Charlie the pig grumps per usual about the lousy state of affairs as he walks past, uninterested. Our vice president and long-term volunteer Chris Seeholzer moves from stall to stall, filling water buckets, whispering sweetly to each animal as she does so. Just feet from Norma Jean, an ancient, blind sheep named Jack is playing with a broom hanging on the wall. I imagine our feathered friend is thinking something along the lines of, "Toto, we're not in Kansas anymore."

~~~

Just three days after Norma Jean joined the Underfoot Family, I walked into the barn to discover her lying right next to Rambo the sheep, her feathered body pressed into his wooly one. Rambo, once the most violent and dangerous animal I've ever known, allowed this—even allowed Norma Jean to peck at his wool, pulling at whatever bits of hay she found interesting. How quickly he became her source of comfort and security, as he has been to so many others who somehow sense his energy, his strength, his empathy.

We call Rambo our guard sheep, but he's so much more than that. For us humans, he's the consummate teacher whose many acts of bravery, wisdom, and empathy demand that we throw out our assumptions about the supposed differences between humans and animals.

The animals unquestionably look to him as leader. Rambo has never sought out friendship. Rather, throughout the years, one animal after another—many sheep, a goat, a duck named Peepers, a potbellied pig, several chickens, and now a turkey—approached him. They need him; there's simply no other way to explain what we witness, for all Rambo does is lay on his deep bed of shavings and hay against one of the barn walls, and the animals come to him. Like so many others, Norma Jean's favorite spot in the barn is wherever Rambo is. In the morning, she scrunches up right next to him, pressing her bulk into his body. When Rambo gets up to stretch his legs or take a stroll, Norma Jean is never far behind. A few years ago, I'd have thought, "What an unusual friendship!" Not anymore. Rambo is always the go-to guy.

～◌

"My tom turkey needs a wife," Patty Reller said when she called us about another adoption. Patty adopted ducks from us several years ago; her small Hollyhock Farm should be renamed "Hollyhock Heaven." Patty completed another adoption application, and a few weeks after she joined us, Norma Jean left the Underfoot Family at CAS to join the Underfoot Family at Patty's place.

In addition to dogs Snorty, Bean, and Hazel, Patty shares her home with four cats, several parrots, an impressive variety of chickens, Flubby the horse, ducks Charlotte and Sir Francis Drake (Charlotte was named after Charlotte Mollo, who serves with her husband Walter McGrath as our dedicated adoption screener), three donkeys, and a delightful goat named Captain Hook.

I enter Hollyhock for a routine checkup after a long drive up a beautiful country road and am greeted by curious animals delighted that someone new is paying them a visit. But trying to take notes as one visits is like taking notes in our barn: Daisy the donkey alternately grabs my pen, my notebook, my sleeve, my hair, and doesn't let go. When I back up to put some distance between us, she follows me. Naturally, I could get out of the field and simply observe from the driveway, but then both Daisy and I would miss out on the fun.

"Norma Jean is a widow," Patty says to me with a straight face, describing how she had lost "both her husbands." But she goes on to speculate that the gentle, unassuming bird is probably happier with Mabel, her female friend, since both her "husbands" were a little aggressive.

It is a cold day when I visit Patty, Norma Jean, and the menagerie who call Hollyhock home. Norma is in her barn under a heat lamp, no more than a foot from her pal Mabel. Like the rest of the birds, cats, and dogs, she's free to roam wherever she pleases on this little piece of heaven. She didn't become much of a lap turkey, as Patty describes her as "a little reserved." Still, this one turkey really is living happily ever after. How I wish their lives mattered to the people three deep at Quattro's, buying "fresh killed birds" the day before Thanksgiving.

~

So many truly satisfying alternatives to turkey exist, and on behalf of these beautiful birds, I entreat you to try one (or lots) this Thanksgiving. My favorite Thanksgiving entrée is a stuffed pumpkin. First, it looks beautiful and festive in the center of the table; second, you can stuff a pumpkin with your favorite holiday stuffing. One year, in fact, volunteer Judi Gelardi and business manager/chef extraordinaire Julie Barone demonstrated three different stuffed pumpkin recipes to a roomful of hungry guests. All were divine!

The vegan community is eager to help you create a meaningful new tradition. Nearly all of us were meat eaters, after all. Many of us, myself included, remember our first turkey-free Thanksgiving. *"What on earth will I cook?"* we wondered, and turned to the Internet. Oh my goodness!! A few gazillion choices, perhaps?

Yes, indeed. Check our Catskill Animal Sanctuary's holiday recipes at casanctuary.org, and Google "vegan Thanksgiving" (or "stuffed pumpkin" if you want to try my idea).

Have fun. And thank you. *Thank you.*

Compassionate Cuisine

During my childhood, dinner was a big event. My mom cycled through perhaps as many as forty main dishes, and I rarely saw her use a recipe. She was quite the cook! Meat, often fairly fancy, was the center of every single dinner. Pork chops. Roast beef. Barbequed chicken. Ribs. Swiss steak. Oyster stew. Spaghetti with meat sauce. Some sort of fancy roast tied up with string. Crab cakes. Vegetable beef soup. We ate food that I can barely make myself type—like leg of lamb, for instance—and occasionally food that I loathed: creamed chipped beef, for instance, whatever *that* was.

On weekends, my dad grilled burgers and steaks that came from cattle raised on our horse farm and slaughtered in Richmond. Mom also filled our plates with vegetables, but during the 1960s and 1970s, with the exception of mouth-watering, award-winning tomatoes grown in Hanover County, Virginia, frozen vegetables were what she—and all other moms I knew—served: Green Giant broccoli spears or Green Giant spinach or Green Giant corn. Asparagus came from a can; salads were made with iceberg lettuce.

"Two bites" was the rule at our table. My siblings and I had to eat two bites of every food item on our plate before we could "be excused." It was always the vegetable, never the meat, that we shunned.

Once I moved to Boston in the 1980s, I mostly eschewed beef. As (largely) unconscious as I was about what food animals endure as they are grown and processed to feed humans, some choices always felt wrong. For instance, I've eaten lobster a grand total of once. The idea of boiling a living being to death always seemed beyond the pale to me. And *veal?* Before I'd even heard of the Humane Farming Association and the veal boycott that resulted in veal consumption plummeting from almost 3.5 million slaughtered in 1986 to fewer than one million in 2011, I knew how veal calves were raised. HFA's outreach efforts were truly far-reaching! The point is that, even though I've always been somewhat health conscious, my movement away from one meat at a time, beginning with beef, came from concern for the animals.

Still, I loved tuna, swordfish, shrimp, and chicken. Far from being a chef, I liked to cook, had eight or ten favorite recipes, and my partner Jesse and I could throw dinner parties that kept friends coming back . . . even if mostly for the wine! We ate chicken at least once a week, ditto with seafood, and every now and then, I'd make an exception to my "no beef" rule to enjoy a very occasional burger at Doyle's, a favorite Jamaica Plain hangout. I'd have been called a "flexetarian" in my twenties and early thirties, if the term had existed.

I became vegetarian around age thirty-five. The trouble was that as I eliminated chicken, then fish, then shellfish, I wasn't replacing those foods with enough good vegetarian dishes. My vegetarian dinners were mostly stir-fries, baked potatoes stuffed with sautéed vegetables, and salads—and Jesse made a rockin' chili. My diet was healthy, but somewhat monotonous.

I was really in trouble when I decided to go vegan, mostly because processed foods like soy, tempeh, and seitan didn't appeal to me. I needed more variety than I could produce myself, so I turned to cookbooks. They helped. Throughout the years, my limited repertoire has expanded to include wonderful soups and stews, variations on rice and beans, a few fun things with roasted root vegetables . . . and lots of fancy salads. I can follow a recipe—even enhance it—and create a nice meal, but I'm far from being a confident or accomplished cook.

None of this is to say that veganism is difficult for me. Quite the opposite! It is a joyful, healthy, and life-affirming experience. It is a *delicious* experience, whether I create a favorite recipe from *Crazy Sexy Kitchen, Big Vegan, The Inspired Vegan, Veganomicon, The Vegan Table,* or others; eat at vegan restaurants or at the homes of friends who are better cooks than I; or sample our chef's inspired cooking. At other times, I'm content with meals that my friend Lucile would call "perfectly fine." As a vegan, in other words, I have the same level of cooking skill that I did as a meat eater (though my goal in 2013 is to cook as well as my mom did, only with different ingredients).

I bring up my food history not to be discouraging, but to be honest: *I get it.*

I get that for plenty of people, the idea of eating fruits and vegetables, grains, legumes, and nuts feels restrictive. I get that for some of you, the prospect of going vegan feels overwhelming: where, and how, do you begin? I get that it might even feel a little *nuts,* no pun intended! (Keep in mind that I've been having this conversation with meat eaters for a while now, so not only do I remember my own struggles, but I constantly hear about other folks' challenges.)

Food is a complex issue. Culture, tradition, public policy, personal identity, family history, habit, economics, geography, health, the quality and availability of good produce, and so much more all influence our food choices. For many of us, what we cook feels like *who we are.* I'll never forget the time a woman told me that asking her to give up meat and dairy almost felt like asking her to give up her children. She laughed when she said it, but everyone in the room got her point. Considering the possibility of substituting lentil walnut loaf, vegan enchiladas, and ginger sesame tofu stir-fry for spaghetti and meatballs, lasagna, and pot roast was terrifying. For those of us who have the luxury of having enough food to eat, our food is often a comfort—and comfort matters. I get it.

But as much or more as any of the above, *brainwashing* influences our food choices. For the past several decades, agribusiness giants, powerful lobbies like the National Cattlemen's Beef Association and the Dairy Council, their many friends in Congress, and the USDA have coalesced to convince us that we need meat and dairy to be healthy. Partly because they've also kept prices artificially low, Americans of every stripe have embraced their message. We currently consume more meat and dairy than any country in the world (just 5 percent of the world's population, we consume almost 17 percent of the meat), with consequences ranging from shameful national health crises to environmental disasters that were incomprehensible just decades ago.

But the tide is turning, folks. As you read this book, I invite you to consider coming along for the ride. There are so very many good reasons to do so.

I'll never forget the day we elected Barack Obama as President of the United States. As a former southerner exposed to ugly racism throughout my childhood, I wept with joy. *We had done it.* Racism was not erased with this historic victory, of course, but the victory highlighted two great truths about our country. First, America is remarkably adept at rapid paradigm shifts. Second, as we bump along through history, we become more enlightened, not less. Social change is a tall order. Yet despite the noisy few who cling to archaic ideas that fail to serve the greater good, most of us march toward a more enlightened view of the world.

I feel another paradigm shift happening in our country right now as momentous as Obama's election. Faster than anyone anticipated, and for all the right reasons, Americans are changing their eating habits. Meat eating is falling out of fashion. According to the USDA, consumption took a nose dive from 2007 to 2012, dropping more than 12 percent. *That's huge.*

It's easy, of course, to understand why. Firstly, *everyone* other than the industry is talking about the disastrous consequences of our dietary choices. I've already mentioned some of the books and documentaries that have brought brutal practices to light and revealed their impact on the planet. But there's much more happening. Celebrities are going vegan and talking about it, vegan cookbooks like my friend Kris Carr's *Crazy Sexy Kitchen* are at the top of bestseller lists, and wonderful organizations like Mercy For Animals are releasing exposés of what goes on inside factory farms. Further, Americans are fatter and sicker than we have ever been, and doctors who've long advocated vegan diets for health reasons—folks like Neal Barnard, Caldwell Esselstyn, Michael Greger, and John MacDougall—are getting lots of air time. You'd sort of have to live in a box in 2013 to be unaware of the impact of our food choices on animals, on our health, and on the planet. And if you do? Chances are that box has either been parched or drenched. In 2012 alone, most of our nation was rocked either by horrific hurricanes and flooding or by drought. Even most conservatives, invested though they are in maintaining the status quo, are acknowledging that global warming is real. Its primary cause, of course, is our diet.

I have a cynical friend who says "people don't change." I always respond with, "You're wrong." Most human beings don't want to hurt animals. They don't want their children to have diabetes. And I seriously don't think they want to be responsible for the destruction of the planet. We've become far more aware of the devastation wrought by agribusiness in the last few years, and we don't like what we see. The vegan train is leaving the station, friends. Catch a ride!

~

Even after all these years of knowing my diet, my dad will still sometimes joke, "What are you having for dinner: sticks and leaves, or nuts and grass?" And on many of our tours, someone with a stricken look will ask: "But what do you *eat?*" It's been clear for many years that people transitioning to a vegan diet—or even those who want to begin by reducing their meat and dairy consumption—need support.

And so, in 2010, Compassionate Cuisine, Catskill Animal Sanctuary's vegan cooking program, was born. Our belief was that if we're asking people to be part of this exciting and necessary paradigm shift, we should do our part to show them how.

To that end, Compassionate Cuisine, headed by Chef Linda Soper-Kolton, offers a range of small, hands-on classes to meet widely varying needs. Plenty of her students are meat-eaters either interested in reducing their meat consumption or in transitioning to a vegetarian diet. These folks desperately need support! But vegetarians and vegans also sign up: They want to add diversity to their meals, or improve their baking, or what have you. Her classes are varied, ranging from Vegan 101 to baking to Indian cooking to Vegan Thanksgiving. Even better, they are small (ten is a full house), intimate, and fun. As Linda says, "I want people to leave our classes thinking, 'That was really fun and I can do this at home!' Cooking vegan is not about what you *can't* eat, but all the new and different things you *can* eat that you may never explore."

Let's take a peek.

I tiptoe through the front door and into the dining room, taking a seat at the huge farm table built by Caleb's dad. Because Caleb opened up the wall that once separated the dining room from the kitchen, I have a full view of all the goings on in the teaching kitchen; Linda's Vegan 101 class is in full swing. It's the class for newbies and wannabes, and Linda is teaching people to make three quick, simple dishes: macaroni and cheese, tempeh and peppers, and veggie burgers.

"Hi!" I whisper to Linda, her volunteer assistants Helen and Lois, and twelve students sitting along the outside of the stainless steel tables arranged in an L. The hands-on class is overbooked; they often are.

Helen walks over and places a slice of bread with red pepper hummus on a napkin beside me. "Such a quiet group tonight," she whispers. "Last night's group for the Mediterranean class was a wild bunch!"

The group does seem both earnest and intent. Two women take notes as Linda speaks, but mostly, students watch as she glides between the stove, where she gives some beans a quick stir, and the tables, where she offers a recommendation for how to slice on the diagonal or how full to fill a food processor.

"There's no reason to fear these things," she explains, pointing to the food processor. "They only have two buttons—'on' and 'pulse.' Just be sure not to blend the crap out of your foods—if you become friends with the pulse button, you'll keep the color and different consistencies of your foods."

Among the students are a family staying at the Homestead—a ten-year-old named Lauren, her mom, her aunt, and her grand-daughter; a young pregnant woman and her husband; a retired couple; and two CAS volunteers. Louise Brinkerhoff, a senior at Bard College, has volunteered with us for three years. Lisa Cutten, our county's Deputy Budget Director, is a regular in the barn on Saturdays.

The kitchen smells like a kitchen should. The fireplace mantle is lined from one end to the other with spices stacked two rows deep: on the bottom, ball jars filled with spices; on top, Frontier brand organic spices. It is a study in contrasts: the room itself is made of brick, has wide plank flooring and a fireplace, but the appliances and work tables are all stainless steel. We were advised that no other material made sense in a teaching kitchen, and now that I'm watching Linda's class, I'm glad we listened.

Linda pairs husband Dan and wife Allie against each other in a "who can make the fastest burgers" contest. "You're going to be at a little of a disadvantage," she says to pregnant Allie, handing her a "regular spatula" as opposed to Dan's "offset one."

A woman in her thirties comments that she'd do a lot more vegetarian cooking if it weren't so much work.

"Thanks for that comment!" Linda says. "Cooking is clearly one of the challenging things for folks who are trying to move to a vegan diet—kinda funny, right? But it's true! You come home from a long day and feel overwhelmed: there's no way in heck you're gonna chop vegetables for an hour before you start to cook."

"Exactly!" the student says. "I work and I have two young kids."

"So you prep in advance," Linda suggests. "If you make sure that your essentials are always prepped—your grain, your bean, your fresh diced vegetables—then your meals are a breeze." Using one of tonight's dishes as an example, she continues: "So, with the tempeh and peppers dish, dice up the peppers and tempeh on Sunday afternoon and throw them in a ziplock bag so that when you come home on Monday afternoon, it's easy to put your meal together."

"Peppers freeze well, too—they don't fall apart like zucchini or similar vegetables do," Helen offers.

Linda turns to check on young Lauren, who is diligently stirring the mac and cheese.

"How are you doing?" Linda asks.

"Fine," Lauren whispers as she takes both the pre-measured mustard and miso to add to the pot.

A young woman asks about nutritional yeast.

"Aaah . . . another good question," Linda responds, and offers a simple definition. "For vegans, it's super-duper important because it's one of the best sources of B12. I only recently found out I was severely deficient in B12. So . . . sea vegetables have B12, and so does nutritional yeast. Sprinkle it on soups, in salads, on popcorn—it has a cheesy flavor. Just when you're cooking, adjust your salt, because it tastes a little salty."

A tall, bearded man in his late twenties pops behind the counter to help with the veggie burgers.

Students talk briefly about the grains they use: millet, quinoa, barley, amaranth; two acknowledge that they've never ventured farther than brown rice. "Look at the bulk bins," Linda suggests. "Try something new every week—they're basically all prepared the same way . . . and if you soak them ahead of time, they're much more digestible."

Linda glides around the kitchen, looking from student to student and pointing to one dish after another, updating the audience on how close dinner is to being ready, interspersing one nutritional tip after another with self-deprecating jokes.

Bard College student Louise volunteers to brown the seitan for the seitan and peppers dish. This is not a simple task; the skillets are packed. A few minutes later, the timer for the burgers goes off. "Voila!" Linda exclaims, then reminds folks to always turn the oven off and let them sit for a few minutes.

"Otherwise, they'll fall apart and you'll be eating veggie crumbles instead of a burger."

As the macaroni and cheese comes out of the oven, a student passes out utensils. She walks over to me. "You gonna have some food?" she asks.

"That's the *real* reason I'm here," I respond. "Taking notes for the book is just a ploy to get in the door."

While the burgers do their thing in the oven, Linda dishes out macaroni and cheese and tempeh and peppers for her students. Lauren dives into the mac and cheese.

"How do you like it?" Linda asks.

"It is really good and I don't even *like* squash," Lauren replies.

From where I sit, twelve students have gotten more than they bargained for in a two-hour Vegan 101 class. They have learned to make three vegan entrées that are quick, healthy, delicious, and affordable. Secondly, they've gotten dinner! Next, they've enjoyed the camaraderie of like-minded people, since everyone here aspires to lessen their footprints and to eat more kindly. Finally, thanks to Linda's freely-offered nutritional tips, they have information that can instantly improve the health value and the taste of many of their meals.

As student Faith Sprague commented, "Linda gives me so many ideas of things to eat, how to prepare a food like tofu . . . even simple tips like freezing burgers: who knew?"

Lauren's aunt, a lovely woman named Theresa, concurs: "Linda showed us that veganism is such a creative way to approach food. It's *the opposite* of limiting; it busts your paradigm right open."

During her two-hour Vegan 101 class, Linda has sung the praises of nutritional yeast, sea vegetables, dehydrated vegetables, parchment paper, turmeric, miso, tahini, and more. My guess is that twelve students are leaving the class singing *hers*. "People used to think that vegan food was weird and tasteless and uninspired," she says. "But they're learning that *it really is possible* to eat compassion-ately and to *enjoy* it!"

Yes they are, Linda, thanks to you!

Linda and I invite you to join us for the next Vegan 101, or for something far more fancy. Meantime, she has generously agreed to share her recipes for veggie burgers, macaroni and cheese, and tempeh and peppers.

Veggie Burgers & Roasted Sweet Potato Wedges
Serves 6-8

Ingredients

2 cups cooked beans or lentils of your choice

1 cup cooked grain of your choice (millet, quinoa, etc.)

1 tablespoon olive oil

½ cup onion, chopped

2 cloves garlic, peeled

1 large carrot, peeled and grated (you can use a small sweet potato instead)

½ teaspoon ground cumin

1 cup minced kale

2 tablespoons ketchup

1 tablespoon nutritional yeast

½ teaspoon salt

½ cup sunflower seeds, pulsed in a food processor

Directions

1. Preheat the oven to 375 degrees F. Line two baking sheets with parchment paper.

2. In a food processor, add all but ¼ cup of the beans and the cooked grains and pulse until blended, but still chunky.

3. Add in the onion, garlic, 1 tablespoon olive oil, carrot, cumin, kale, ketchup, nutritional yeast flakes, sunflower seeds, and salt to the bean mixture and pulse until combined. Finally, add the ¼ of reserved beans and combine so the beans are still visible and chunky.

4. Lightly oil one of the baking trays lined with parchment. Form mixture into patties and place them on the prepared baking sheet. Use an ice cream scoop for ease and uniformity. Press with a spatula to flatten.

5. Place the burgers into the oven and bake for 20–25 minutes, until dry on the top and slightly firm. Remove from oven and let set for 5 minutes before removing from tray.

6. Serve with or without buns along with some fries. Add lettuce, tomatoes, or condiments of your choice.

Mac & "Cheeze"
Serves 8

Ingredients

2 cups butternut squash (or any orange squash), peeled and cut into small
    pieces
1½ cups coconut milk
1 cup water
2–4 garlic cloves, minced
½ teaspoon ground nutmeg
1 tablespoon onion powder
1 tablespoon organic mellow white miso
2 tablespoons organic tamari (soy sauce)
1 teaspoon Dijon mustard
1½ teaspoons salt, divided
2 teaspoons potato starch (or organic cornstarch)
1 pound pasta
1 cup Brazil nuts
½ cup nutritional yeast, divided

Directions

1. Preheat the oven to 375 degrees F. Place a medium pot of salted water on the stove to boil.

2. In a medium saucepan, cook the butternut squash in the coconut milk and water over medium heat until soft. Add water to cover the vegetables if necessary.

3. Add 2 cloves of garlic, nutmeg, miso, tamari, mustard, ¼ cup of the nutritional yeast and salt and cook gently for another 5 minutes. Add the potato starch to thicken.

4. Remove from heat and cool slightly.

5. While the mixture is cooling, boil pasta in water until al dente, according to package instructions. Drain and pour into a medium mixing bowl.

6. Make the Brazil nut parmesan: Place Brazil nuts, ¼ cup nutritional yeast, ½ teaspoon of salt, and garlic in a food processor and pulse until blended and the consistency of breadcrumbs is achieved. Season to taste with more nutritional yeast or salt.

7. Carefully pour the slightly cooled squash mixture into a blender or food processor and blend until creamy.

8. Pour the mixture over the pasta and stir to coat.

9. Pour the pasta into a baking pan and sprinkle liberally with the Brazil nut parmesan.

10. Bake in the oven for 20–25 minutes or until the top is slightly browned. Serve.

Tempeh & Peppers
Serves 4

Ingredients

1½ cups water

½ cup organic tamari (soy sauce)

1 bay leaf

1 package organic tempeh, cut into ¼ inch strips

¼ cup extra virgin olive oil

2 bell peppers, seeded and stemmed and cut into strips (green, red, yellow, or mix them up)

2 yellow onions, halved and sliced thin

1 teaspoon kosher salt

1 teaspoon freshly ground black pepper

1 teaspoon dried oregano

2 teaspoons ground fennel seed

1 teaspoon dried basil

4 garlic cloves, peeled and chopped

¼ teaspoon red pepper flakes

4–6 fresh whole grain sandwich rolls or cooked, warmed brown rice, or grain of your choice (optional)

Hint: Wash, trim, and prepare all the vegetables and herbs while the tempeh is simmering.

Directions

1. In a small pot, combine the tempeh, water, tamari, and bay leaf. Simmer for 15 minutes and remove tempeh from pot.

2. Heat the oil in a heavy large skillet over medium heat. Add the tempeh and brown on both sides. Remove from the pan and set aside on a plate.

3. Keeping the pan over medium heat, add the peppers, onions, salt, and pepper and cook until golden brown and tender, about 20 minutes. Add the oregano, basil, garlic, chili flakes, and fennel and cook 2 more minutes.

4. Add the tempeh back to the pan and stir gently to combine. Let everything simmer for a few minutes so the flavors combine. Serve in bowls alone, with brown rice, or as a sandwich.

If We Can Get Them Out

In the midst of this country's economic downturn, it has been our pleasure and our privilege to receive animals from caring people who reach out for help because they can no longer afford to care for their beloved family members. Farms are in foreclosure; people are losing their rural rental property . . . or simply can't afford the price of hay. They call, generally in tears, when they have run out of options. We help as many as we can. This situation was different.

We were called by another rescue about a woman who "needed help." She had three animals—a mare and two older stallions, all Thoroughbreds—and could not afford them. "They might be in bad shape," the rescuer said. State police were weighing whether to arrest the woman or simply to force her to surrender the animals to avoid prosecution.

Early on a Tuesday morning, I called the woman. She talked about how she'd lost her job, how she could barely afford to put gas in her car, how the horses were forty minutes from her home, how she had to haul hay and water in a broken down vehicle that might just die on any of these trips, and how no one wanted the animals—especially not two older stallions.

"They're Thoroughbred stallions," she said. "Who in their right mind would take them?" I wanted to ask why she didn't neuter them many years ago, but didn't. (In conversations with people who are neglecting their animals but *seem* willing to surrender them, one stays as cool as a cucumber. A comment made in anger or frustration can quickly derail the process.)

We would take the stallions. Despite all the challenges— needing to house them far away from mares, not being able to turn them out with other horses, the expense of gelding, the risks associated with gelding older horses (one is fourteen, the other twenty-one), and the likelihood that they will be extremely hard to handle—we would take them.

I told the owner that we needed to see the boys before they came to Catskill Animal Sanctuary, and that we would find a good home for the mare. Questions loomed. Were they strong enough to make the trip to their new home? A psychologically damaged stallion could be extremely dangerous; what had their imprisonment done to them? What was their physical condition? If they were exceptionally thin and debilitated, we'd first need to put weight on them before they'd be able to withstand being gelded, and that would mean more time juggling the management challenges inherent with stallions.

Allen Landes, who works full time as a hospital biologist at Albany Medical Center but for years was a godsend at CAS every Thursday, lived just three miles from the horses' home. He agreed to take a look at them. I relayed directions from the owner; Allen went right after work. He used his cell to call from what he described as some sort of derelict camp.

"Kathy, this can't be the place," Allen said. "There's no sign of life here. Are you sure you understood the directions?"

"It's the place," I assured him, having gotten a clear description from the horse owner.

Allen walked into a place that looked like it hadn't seen life in decades. It was obvious: The woman had hidden her animals on a long-abandoned property—no boarding fees, no one to complain about the condition of the animals.

"No horses in here," Allen said as he traipsed through one ramshackle shack after another. "Nobody here," he said as he moved through the derelict remains of another. Twice more he asked whether the directions were accurate. Allen was about to give up when he spotted a forlorn-looking barn at the base of a hill at the back of the property. "Ugh," he uttered. "I hope they're not in here."

They *were* "in here."

For many years, Allen served as a board member for an Albany-based rescue and assisted with seizures when owners were arrested for cruelty. The first words out of his mouth were, "God . . . this is the worst situation I've ever seen."

Three horses. All bone thin. Living in darkness in tiny, rat-infested stalls that had never been cleaned. Blanketed, but with blankets that were stuck to their manure-encrusted coats and would have to be cut away from the animals' bodies. Water buckets bone dry. "There's no sign that there's ever been water here," Allen said. One stallion was shivering violently and spent most of the half-hour Allen was there lying down, rats crawling around him. His breath was rapid and shallow.

It turns out that the woman had been arrested for animal cruelty eight years ago, when one of her horses, a Thoroughbred stallion, had to be euthanized—he was simply too far gone. While the first order of business would be to safely remove these horses, we also knew that euthanasia might be the only humane option. We hoped to hell that this would not be the case.

On moving day, Allen and his friend Bob met Corrine the hauler, Sue McDonough—a thirty-year veteran of the state police force and a crackerjack cruelty investigator—and Tina Murray of Harmony Hill Rescue. The horses' owner met them at the property; Sue would determine later whether to file formal charges. For now, the group's only goal was the safe removal of the animals.

The partially collapsed, windowless building that had been their prison for so many years was at the bottom of a steep slope. Corinne had to park some hundred feet from it. The group would walk the horses out to assess their condition, and, they hoped, successfully load them.

"The mare came out first," Allen told us later. And that was when the owner made the first comment that nearly sent him over the edge. Her blanket was not only filthy and encrusted, but it was also so torn and tangled that traveling in it would have been dangerous.

As slowly and carefully as they removed it, however, much of the mare's skin came with it. When it finally came off, the woman eyed the bald and scabby body, some two hundred pounds underweight, and said, "Oh, I thought she was going to be skinny—she looks pretty good!"

"Good *lord*, Allen—way to bite your tongue!" I said to him.

"Yeah, tell me about it," he said. "Gotta keep your eye on the prize."

Cas, the bright red stallion, was rail thin and shivering violently. "He was a wreck," Allen said. "It took two doses of tranquilizer before we could get him on the trailer." Eventually, under the hazy fog of Acepromazine, they did.

It was Noah, the final horse and the one the owner referred to as "her boy," who struggled most. Noah left the barn taking baby steps; it was all he could manage on malformed, painful hooves. Recalling Noah's effort to leave his dark prison, Allen remembers two things: first, how Noah, accustomed to living in total darkness, kept blinking as his eyes adjusted to a sunny day; second, how he kept falling. "He was in so much pain," Allen said, "and he was so, so weak."

Corinne's trailer that day was a "step up" with no ramp for the animals. The trailer floor was a good foot above the ground. The first two horses had managed to step up; Noah simply lacked the physical strength.

"He tried so hard," Allen said. He'd step onto the trailer with his right leg, then fall as he tried to bring his left leg up. Sometimes he'd crumple to the ground; sometimes he'd fall hard. At one point, the owner said, "Maybe we should leave him here." Allen later told us, "That was when I saw steam coming out of Tina's ears."

But the old stallion did not give up, and neither did the intrepid team of caretakers. Eventually, they figured out how to sort of slide him onto the trailer with Tina pulling from the front, the others shoving from the rear, and Noah doing his damndest to help.

They had done it, and were on their way home.

"Where are you?" I asked Corrine when she picked up her cell phone. The mare had been delivered to Harmony Hill; the entire CAS crew had been anxiously awaiting the arrival of the two stallions, and it was now nearly 7 p.m.

"I'm on Old Stage," she said. "It's been a hell of a day."

I hung up and shouted into the kitchen, "They'll be here in two minutes, guys!" Six bodies bundled against the cold appeared.

"What kind of shape are they in?" April asked.

"Corrine says 'awful.'"

A tall, narrow chestnut was the first off the trailer. He hobbled one painful, unsteady step at a time as a chorus of voices welcomed him, praising each step. A torn and tangled horse blanket hung as stiff as cardboard on his narrow body. He was shaking violently—the most violent shiver I've ever seen in an animal. Whether the shiver was from cold or pain or both, we couldn't yet be sure. His eyes were uncertain.

"This is Cas," Allen told us. Evidently it was short for Casanova.

Interesting, I noted to myself . . . his name was the same as ours.

Cas was a true chestnut—a bright rusty color—and as narrow as a reed. He was all bone, no flesh. He rocked from side to side as we cut the fetid blanket from his body, tightening and watching us warily.

"Can you get him something warm?" I asked April.

"What size?" she asked.

"A seventy-four would work fine," I responded, and April returned with a blanket that looked like it could melt an igloo. She held Cas while I slipped the blanket over the pockmarked body. The hair on a quarter of his body was missing. When I rubbed my hand over what remained, it came out in chunks. Cas flinched; I stopped. "Okay, good boy," I praised him. "It's okay. You just eat and get warm."

A darker horse—a rare deep rusty-brown known as liver chestnut—was next. This was Noah. Noah had turned himself sideways during the trip, and now, as Walt attempted to turn him around, he collapsed once again, crashing hard to his knees. Alex, Keefe, April, and I crowded around the outside of the trailer, encouraging him.

"You can do it, Noah," I offered.

"It's okay, boy, you're okay," Allen said, and then to us added, "He's worked really hard to get here."

Walt steadied him and he stood. We watched as, inch by inch, he made his way toward smiling faces. The mere act of walking was an act of bravery. I looked down and understood why. Noah's hooves were grotesquely malformed. His left front foot, in particular, was violently twisted, resembling a wrung-out dishtowel.

Noah had fallen again during the trip. "He fell hard," Allen said. He was spent. Yet still he moved toward us, one tiny, excruciating step at a time, until a few moments later, Noah was in what I hoped to hell felt like heaven: a deeply bedded stall with two buckets of fresh water and a hay rack filled to overflowing with the stuff of life.

When Cas and Noah arrived tonight, they reminded me: *This is why you do this work.* Yes, absolutely. Rescues like this are precious moments at Catskill Animal Sanctuary. They are our moments of glory.

Over the next few weeks, visitors and volunteers meet these two boys. Many cry; many ask, "Aren't you angry?" In fact, judging by how often we're asked the question, many people struggle with why we don't walk around wanting to punch people. But we don't. On most days, in fact, I walk around with a heart so filled with joy that I feel it could burst at any moment. Anger serves neither us nor the animals we are here to help. Joy does. Joy works.

Noah and Cas have only been with us for two weeks, and a woman I'll call Marsha is hounding us to euthanize Noah. Her suggestion is premature at best; groundless, or something far more cynical, at worst, and entirely inappropriate, regardless. She saw Noah the day after his arrival at Catskill Animal Sanctuary, the day after he tried far harder than a lesser spirit would have to free himself from misery. She heard the stories of how he fell, time after time, of how he wouldn't give up. She heard about when he arrived at our barn and fell again, about how he took fifty or more miniscule steps between the trailer and his stall just twelve feet from the door. A horse who walked normally would have covered the distance in five or six steps.

"What are you doing about Noah?" she asks in her third phone call in two weeks.

"We didn't save his life in order to end it," I said to her. "We've got to see how he recovers."

I understand her perspective: She believes Noah is too far gone to have any quality of life, and that putting him down might be the kindest thing to do for him. It would also open a space for another needy animal. She doesn't say this; she knows she doesn't need to, but it's the real reason she's almost *lobbying* for death by lethal injection. She knows I know this. Marsha has been in animal rescue for a very long time; where we disagree is always about euthanasia.

Her "just put him down" statements leave me feeling as though she truly doesn't understand our mission. "Emergency rescue" does not mean "remove animals from wretched conditions, then after that, if you need to euthanize them, it's no big deal." To us, the definition of emergency rescue is something like this: Take the neediest first—the ones who have run out of options. If there is hope, no matter their physical or mental condition, *try mightily* to save them. If they need weight, give them the optimum program for their species, age, and condition. If they need love, let them guide you in how you offer it; many aren't ready for a full-blown love fest. If they need encouragement, offer it in abundance. Only when there is no hope, but instead only suffering, make the choice that they would ask you to make if they could speak. Make it quickly and with confidence—prolonged suffering is not an option—and send them off with all the love you can muster.

It's true that when we look at Noah's gnarled hooves, it's hard to imagine that the coffin bone, the large bone inside the hoof, hasn't rotated. If it *has* rotated, then Noah is permanently crippled, and we'll have to wrestle with the agonizing choice of whether to have him live his entire life in a deeply bedded stall or whether to put him down. Most horses would become stir-crazy and depressed living in a small box; the decision would be unambiguous. I'm not so sure about Noah. He's a mellow fellow.

On the other hand, if by some miracle there is no rotation, will our farrier Corey Hedderman be able to reshape his hooves over time to prevent the constant pain that plagues him? Will Noah ever be able to walk normally? If not normally, then at least without pain? These are the questions being raised while we're in the "wait and see" stage of his recovery. Yes, it may be that we ultimately do have to euthanize him, but while we're waiting to look at hoof and lower-leg X-rays, we are certainly not thinking about euthanasia. He's in pain, but not in agony, and he's far from giving up. Give me a break! Noah's attitude and appetite are superb; he's gained sixty pounds; the hard clumps of crusted manure have been removed from his coat as volunteer after volunteer asks to groom him. For now, at least, he appears perfectly happy to remain in his stall, his head always buried in his hay, while his strength returns.

No. We will not euthanize him.

"This is inappropriate, Marsha," I say before hanging up. "Please don't call me again."

Once again, Noah's head was where it always is: buried in his hay. Even after a night of eating with gusto, I knew he'd be ready for more, so I entered his stall carrying two big flakes of timothy-alfalfa mix.

"Good morning, lovely boy," I said to him as I placed one flake, then two, in the rack on the wall. Noah didn't want his hay in the rack. The old horse wanted it on the floor, which he indicated by lowering his head and looking directly at the ground. So I placed it there, in the corner. Noah tore out a huge mouthful, then lifted his head, looked right into my eyes, and blinked slowly.

Aaah, the blink that says so much, all of it good. You've seen it in your cats. Many animals use it: slow, purposeful, direct. I believe it means everything from "I love you" to "Thanks for the food" to "I'm happy to be here."

I blinked back and said, "You're welcome, big boy. We're so happy you're here." In fact, I was instantly bliss on two legs, for that simple, slow blink told me a great deal. In essence, it was a thank you: an acknowledgement of what we're doing for him.

Few rescued animals communicate so openly and so quickly. In fact for many, eye contact is far too threatening. But not for Noah. Despite tremendous and prolonged suffering, Noah has a big, open, grateful heart and is a communicator of the first order. "My foot hurts," he says, as he lifts his hoof when he knows we're looking, then often turns his head to ensure we see what he's doing. "Can you help me?" Noah speaks with his eyes; he speaks with slow turns of his head to look all the way behind his body. Like the great sheep Rambo, whose story is told in my first book and continues in this one, he's wise enough to know that we're paying attention, wiser still to find ways to tell us what's on his mind.

"Today's a big day, beautiful," I said to him. "We're going for a walk."

The momentous decision to take Noah for short daily walks came the day I entered the barn and saw our farrier Corey standing in the aisle with a huge grin on his face. *Could it be?* I wondered.

"There's no rotation," he said. He'd just examined the X-rays with our vet Heather O'Leary.

"Get *out*!" I shrieked. Rambo startled; Barbie the hen ran for her life. We were all incredulous.

As mangled and deformed as his left front hoof was, we were all but certain that the bone inside that hoof had rotated. But it hadn't. We had our shot. If Corey could work yet another miracle, Noah had a chance at a relatively normal life.

"Come on, big boy," I say as I slide green halter over red head. "We're going for a walk! Can you believe it?"

I take the end of the lead rope and move backward toward the door so that the rope's six-foot length extends between us. I figure it's better for Noah to navigate his own turn. He, not I, knows which joints and hooves hurt and how best to place them to minimize the pain.

At the open stall door, he stretches his head out, turning left and looking down the long aisle, turning right as Millie the potbelly pig trots past in her relentless search for food. I don't rush the process. Let each animal heal in his own way, on his own terms. That's our way. While Noah's reluctance to leave could be *fear-based*—his entire world prior to coming here was a dark, windowless hovel—I strongly suspect this behavior is *all* about a reluctance to leave his hay. Like many chronically starved animals,

Noah is obsessed with food. Maybe he needs a moment to get comfortable with leaving it.

It's been three or four minutes of standing, looking, but here he comes toward me now. I want him to see the encouragement in my face, and I want to be able to watch how he's moving. So I walk backward, facing him. Noah moves cautiously, and in slow motion. He steps tentatively with his front feet; I notice his back legs drop dramatically at the fetlock joint—he has the equine equivalent of extremely weak ankles. Noah walks about thirty feet. Then he stops, looks at me, turns back, and looks at his stall.

"Okay, babe, enough for today? Thanks for telling me," I praise my friend. We circle as widely as we can, and take more small, cautious steps back to the comfort of his stall. Within seconds, Noah's head is buried deeply in his hay.

Ten days later, Noah is walking the full length of our 120-foot barn. There have been two scares: once his left front leg buckled and he pitched forward, and once his back end gave out and he crashed to the ground. But if these moments fazed him, he didn't let on. Both times he got up unassisted, literally and figuratively shook himself off, and continued as if he understood that an occasional slip up was part of the recovery process.

Today, Norma Jean the turkey is all feathered bliss as she naps in the aisle. Noah lowers his head in a gentle greeting that Norma Jean knows she needn't fear. So, too, with Hannah the sheep, marching through the hay room as we walk past. She stops, lifts her head to Noah in confident greeting, and sheep and horse stand nose-to-nose for a moment. When she moves on, we do, too.

Noah seems to have gotten his sea legs—his steps are much more confident than they were ten days ago, though they're still a little cautious. He apparently knows that one could give out at any time. Noah delights us as he moves from one curiosity to the next. Sometimes it's a broom or a wheelbarrow filled with shavings, but far more often it's a living thing—a chicken or turkey, a sheep or pig.

Hazel the adolescent piglet trots up and lifts her pink snout in greeting. Many horses despise pigs; not this one. Noah lowers his head, and as the thousand-pound horse and fifty-pound pig greet each other, all soft breath and innocence, time stops. All is right in this world.

The return trip to Noah's stall is uneventful, except for the fact that the animal we weren't sure would live is pulling me quickly down the aisle. I know why, of course. We turn into his stall, and in an instant, his red head is once again eyeball-deep in the green hay.

A year after Noah and Cas were rescued, Marsha, the woman who'd wanted us to euthanize Noah, stopped in to say hello and to ask if we had room for another blind horse who was about to be seized in a cruelty case. It was just before feeding time. We hadn't spoken since I asked her not to call me back.

"Whatever happened to that horse?" she asked.

"Noah, you mean?" I responded.

"Yeah . . . Noah. Did he make it?"

I led Marsha out into large pasture behind the barn. "Animals!" I called to the group grazing a couple hundred feet from us. "Bowie, Noah, Hazelnut! Crystal! Come on, guys! It's dinnertime!"

Five heads looked up, and Noah, the undisputed head of the herd, came galloping toward us. His head was high and he was running, pain-free, leading his friends to the barn for dinner. These days, in fact, Noah spends a great deal of time running—in part, because he's a Thoroughbred, and Thoroughbreds truly like to run, but I think also, in part, *because he can.*

Why It Matters

"How do you guys do it?" people often ask. What they mean is, "How do you not walk around in a constant stage of rage, given what you witness?" *Specifically* what they mean is, "Don't you want to hurt people who starve their animals?" They rarely, if ever, mean, "Wow—isn't it tough to know what you know about food production and animal suffering and not go off the deep end?"

I usually respond (depending on the asker) with a more elaborate combination of the following:

1. Rage serves neither us nor the animals.
2. It's impossible to be around animals and feel angry. They're the perfect antidote. Besides, there's no time or room for anger. All one's energy and attention goes to caring for the needy beings right in front of you.
3. The suffering of animals at the hands of "abusers"—people who starve them, for instance—pales next to the suffering of those whom we eat, wear, or otherwise use for our purposes.

You, good person, are the cause of more suffering than the suffering that angers you.

It's really *not* the work involved in caring for neglected animals that makes our work emotionally challenging. For sure, we've accepted animals from situations that cause us to question what it means to be human. But if you surveyed people who run animal rescue organizations, I bet they'd agree with me—other stuff is far more difficult. When a sanctuary rescues a large group of starving animals, all its energy and focus goes to healing those broken beings. These aren't empty words; there truly is no time or room for anger. It's other stuff that will knock you down: law enforcement's collective turning of a blind eye to cases of enormous suffering, failures within the justice system to protect farm animals, and the reality of saying goodbye to beloved friends over and over and over again. While we have gotten "good at death"—our final goodbyes are moments of rare beauty and power, and we very quickly resume business as usual—they also leave us feeling raw, exposed, wrung out. This is the stuff that's bone-crushing. That plus a generous sprinkling of cer-ti-*FI*-a-bly crazy people that our work attracts.

In addition, challenging the status quo ain't always a walk in the park. I love speaking about the work of Catskill Animal Sanctuary and the imperative of veganism. But living in a world that reveres "companion animals" and disregards all the rest is tough for people who know that *all* animals want their lives as much as we want ours, and that suffering feels the same to them as it would to us. To constantly consider how to hone our message so that good people with mainstream beliefs will open their minds and hearts is work that matters, but hell yeah . . . it's tough.

So join us in these pages to look at the work of Catskill Animal Sanctuary through a very different lens.

A Bucket of Need with a Hole in the Bottom

Animal Hoarder: an individual who obsessively accumulates an enormous number of animals, fails to provide minimal nutrition, space, sanitation and medical care, and lives in denial regarding the deteriorating condition of the animals and their surroundings. Hoarders often justify their behavior with irrational and/or nonsensical arguments, and tend to be secretive about their behavior. The recidivism rate for hoarding is thought to be virtually 100 percent.

"Where do your animals come from?" people ask. We didn't always keep this particular record, so it's impossible to speak in terms of exact numbers. Still, the question is easy to respond to in broad terms, and the answer is never what folks expect. It wasn't what we expected, either, when we first began this work.

In ascending order, animals arrive from the following situations:

1.   A tiny percentage of all the animals welcomed to CAS come from situations one thinks of when she hears the words "animal cruelty"—rabbits left to die in Tupperware containers, a single horse chained to a tire and starving to death, etc. Of the 2,500 animals we've accepted, no more than ten have come from situations involving intentional, depraved cruelty such as the above.

2.   Another small percentage of our animals are surrendered by family members of people who have fallen ill or died.

3.   A larger number arrive as a result of economic crisis. In the worst case, people's farms are in foreclosure; in other cases, however, people need to simplify to make ends meet. "It's tough having to choose between feeding your animals and feeding your children," one woman said to us.

4.   Hundreds have come from random situations that one could never even conjure up. Several years ago, for instance, we accepted forty-two chickens from Kansas City, Missouri after a raid on a crystal meth lab where they were living.

5.   Probably as many as 20 percent come from the five boroughs of New York City. These animals are lucky escapees found walking down the streets, having broken free of live poultry, goat, and sheep markets. Others have been found in garbage cans, dumpsters, cemeteries, tied to trees in Central Park, or hiding under cars. We've taken chicks found in a school dumpster (who were presumably thrown away after a classroom egg-hatching project), rabbits in a box placed beside a dumpster, a domestic duck crossing the street in a Brooklyn neighborhood, and a rooster stuffed in a mailbox.

6. Even more arrive from what Dr. Gary Patronek, VP for Animal Welfare and New Program Development at the Animal Rescue League of Boston and a leading expert on the subject, calls "pseudo-sanctuaries"—places that label themselves shelters or sanctuaries. These places sometimes have non-profit status but are run by animal hoarders. Patronek notes that "it's pretty clever to call yourself a sanctuary when you know you're hoarding animals. It sure does make you more palatable to the public." We have accepted animals from ten or more places such as this. A very few reach out on their own; most, however, have come to the attention of law enforcement. Many have been shut down.

7. By far the greatest number of animals accepted by CAS have come from animal hoarders. Sometimes the animals have been seized by the police and cruelty cases ensue against the hoarders. At other times, hoarders have been pressured to surrender their animals to avoid prosecution. By now we know what to expect: The animals arrive filthy, riddled with internal and external parasites, often with their hair coming out in big hunks, sometimes only slightly underweight but sometimes verging on collapse, and often with crippling hoof neglect. Just as often, they arrive as feral animals, never having had attention from a caring human.

Here's the story of one hoarding case. I have changed the names of the people involved; all other details are true.

In the dead of winter 2004, Allison Scott, director of an animal shelter in upstate New York, asked if we could assist with a hoarding case involving domestic and farm animals. Animals from dogs and cats to goats, sheep, llamas, and horses were being seized. We agreed to take the sheep, donkeys, and horses, and headed out early one Saturday morning with a caravan of volunteer animal transporters.

I'll never forget what I saw in the first few minutes at the hoarder's "farm." I'd never been inside the home of a hoarder, so when I walked into the old farmhouse, the words *"This is what they mean"* came to me. "Squalor" is always the term used to describe animal hoarders' homes; squalor was what I was seeing. Huge sunken holes in the floor. Holes in the walls as if someone had used a sledgehammer to bash them in. Inches of dried feces everywhere my eye landed: on floors, stairs, boxes, tabletops, counters. Urine and feces-stained newspaper everywhere. Most of the windows in the old house were non-existent or broken—on a frigid day with a foot of snow on the ground, it was as cold inside the house as outside. Before we arrived, a dead dog hanging by the collar from a light fixture next to a staircase had been removed by shelter employees. The dog had choked to death; the hoarder had left his body dangling, rotting.

"I couldn't deal with it," she said.

Outside in the snow, a thirty-gallon drum contained the remains of cats and chickens, their carcasses burned, perhaps to hide evidence of their mistreatment. Three grown llamas lay dead in various barns, their necks and abdomens ripped wide open. Earlier in the week, investigators saw two starving dogs eating one of them alive.

Once law-enforcement officers got their bearings, it was time to catch the horses. There were no sheep on the property; we were certain they'd been sold for slaughter. Virtually all the equines were feral, having lived without any handling whatsoever; this wasn't going to be easy.

"Could we have some grain please?" I asked the hoarder, figuring we'd need it to entice the frightened animals close enough to toss a lead rope around their necks, hold them, and quickly slip a halter over their heads.

She stared at us blankly. "I don't have grain," she said. Twenty-two equines on the property in weather hovering just above zero degrees . . . and there was no grain. In fact, as far as I could tell, the only food on the entire property for twenty-two horses and untold numbers of other animals was a few bales of hay brought to the farm by investigators.

The hoarder's record dated back to the 1970s when she was arrested for having scores of dogs in her city apartment. Investigators had been to her farm numerous times since the late 1980s, and always found misery—specifically, dead and dying animals. Today, we see the usual: a donkey missing most of his ear, likely due to frostbite; two horses enclosed in a tiny pen with hooves so twisted they may not be able to hobble to the trailer; a dog clearly desperate for food who is a gnarled mass of hair, brambles, and feces. Dead bodies everywhere.

"Do you have any regrets about what happened at all?" a reporter for the local paper asked her.

"Regrets about what? About what?" the woman replied. "They got me for a dog needing a haircut. What else can I say?"

With a crew of expert horse handlers, and some horses who clearly wanted to be *anywhere* but where they were, catching and loading twenty-two equines went relatively quickly. The hoarder stood in the snow the entire time, naming each one as we led it from the property and onto a trailer. "That's Bootsie," she said of a little gray mare with a belly full of worms whose hair was coming out in hunks. "She's Moonshine's daughter."

It was nighttime when we arrived back at Catskill Animal Sanctuary. We were all chilled to the bone. We were all wrung out.

We were all elated. A welcome committee of a dozen or more staff and volunteers were on hand to meet twenty-two full-sized and miniature horses and donkeys who were led, one by one, into a barn filled with warmth, food, and love.

All but one, a gentle chestnut horse named Spirit, survived. Sadly, Spirit was already in organ failure when he arrived at CAS. His friend Freedom, approaching thirty years old, lives with us today.

~~✒~~

It's no exaggeration to say that the problem of animal hoarding is an epidemic in the United States. While about 2,000 new cases are brought to light each year, I'd be willing to bet, based on our experience, that those that never see the light of day outnumber those that do. Let me be more specific: those that are ignored outnumber those in which law enforcement intervenes in an effort to help the animals.

In our region of New York State, for instance, CAS is aware of five hoarders who continue to take in animals, collectively subjecting hundreds to horrific suffering as law enforcement appears to turn a blind eye. While there may, sometimes, be justifiable reasons for what seems to an observer to be gross negligence of duty, the result is always the same: Across the country, hundreds of thousands of animals suffer—and many die—in squalid surroundings, devoid of adequate food and water, while one by one, hoarders insist that nothing is wrong.

Yet *everything* is wrong with animal hoarders. Consider this additional case:

Shortly after Catskill Animal Sanctuary opened in 2001, we were alerted about eighteen animals—two cows, two goats, and fourteen sheep—who were locked in a filthy, rat-infested stall and being fed moldy bagels. A decomposing cow carcass was in the stall;

our cow Molly, now thirteen years old, was her bereft calf. Outside the fetid barn, dogs were chained to a fence without food, water, or shelter. While the hoarder was forced to surrender her animals, she immediately began collecting animals again. Several years later, we accepted chickens and ducks after she was arrested for selling them in a supermarket parking lot. In May 2012, we accepted a horse from her property who'd been living in the woods on a heavy chain without access to water or shelter. I drove by the property often to check on this little horse and to document the neglect. "Cricket" was often tangled in the chain and physically unable to move; she was *always* thirsty. She survived Hurricane Irene as trees swirled and crashed around her, but came to CAS with a permanent limp and a deep scar around her leg from the chain. This local pariah continues hoarding cats, dogs, rabbits, goats, turkeys, chickens, ducks, geese, and an occasional horse or donkey, and is a source of enormous frustration for our local SPCA.

Despite their claims of innocence, animal hoarders cause far more injury, suffering, and death to animals than intentional abusers. And not only do the animals suffer, but people do, too. Hoarding jeopardizes the health, safety, and well-being of any humans living under the same conditions. And yet it's a problem that largely goes unaddressed.

While most of us think that hoarders collect only dogs and cats, this is not the case. Granted, it's relatively easy to hide cats, and cats are readily available, so a high proportion of hoarders really are "crazy cat ladies." Still, hoarders accumulate a variety of animals, from dogs and cats to farm animals and exotics. A hoarder less than two miles from us is reported to have more than one hundred dogs and cats in his house, but he also has dozens of ducks and geese, along with a number of goats, jammed into tiny pens in full view of passersby.

According to the Hoarding of Animals Research Consortium, a group of experts who collaborated for a decade to explore animal hoarding and to find ways to address it, hoarders typically fall into three categories.

- *The Overwhelmed Caregiver* initially provides adequate care for animals they truly care for. Though they may minimize it, they understand on some level that a problem has developed. This type of hoarder is most amenable to receiving help.
- *The Rescuer Hoarder* has a compulsive need to rescue animals from possible death or euthanasia. These folks purposefully acquire animals by trolling the internet, etc., believe no one else can adequately care for them, and rarely, if ever, refuse to accept an animal.
- *The Exploiter Hoarder* is indifferent to the harm caused to animals and acquires them only to serve her own needs. These people may be charming and articulate, have an unhealthy need for control, and are generally considered extremely manipulative. They actively acquire animals, actively reject outside help, and have well-defined plans to evade the law.

I'm not sure what it says about the model that Catskill Animal Sanctuary has never encountered an "overwhelmed caregiver." I see how it's useful to distinguish among companion animal hoarders, but from our vantage point, hoarders of horses and farm animals don't ever seem to be "overwhelmed caregivers" or "rescuer hoarders." Certainly the ones we've dealt with all appear to be exploiters—folks whom Gary Patronek says are "so toxic that the only thing that you can do is to mount an aggressive prosecution."

An aggressive prosecution, however, is never a slam dunk. One would think that photos and videos of filthy, sick, starving animals, photos of dead animals, and expert veterinary testimony would ensure victory. But they don't. And even when cases are successfully prosecuted, the typical sentence is a short period of probation, after which the hoarder begins collecting once again.

The problem is a knotty one. First of all, as Patronek pointed out to me, cruelty laws were written to target *the abuser:* the person who beats his horse to death or throws her cat out the window. No one envisioned hoarding when the laws were written. Second, as they stand now, cruelty laws in every state pertain to the treatment of *an individual animal.* Consequently, the prosecution in hoarding cases must prove that each animal has been mistreated to remove it from the situation. Patronek believes that we need to be able to prove a situation *in toto* by focusing on the entire environment. "When things are this grotesque, the situation should speak for itself," he urges.

The idea of evaluating *competence* to care for animals is a second compelling idea. "It's not a novel legal concept to evaluate people's competence," Patronek points out. The child protection system, for instance, recognizes that a specific skill set is required to adequately care for children. There are ways to go into the home, to try to keep the family intact by providing resources, but then if necessary to take action up to and including removing the child from her incompetent parent. Patronek says, "I don't see why we can't use this model for animals."

Meantime, as experts and lobbying groups work to reform existing law, legitimate shelters and sanctuaries like CAS are on the receiving end of the problem, one that Patronek describes as a "bucket of need with a hole in the bottom."

The Fabulous Four

W hen we learned of four newborn calves desperately needing sanctuary, it took farm manager Kathy Keefe, animal care director Jenn Mackey, and me a nanosecond to say, "Yes . . . we'll take them."

Mind you, we needed four little bulls like we needed holes in our heads. For starters, we already had too many steers, a fact that made for all kinds of housing and pasturing challenges. (Our four cattle "herds" are permanent groupings of animals who like each other. These four babies would be a fifth; that's a lot of pasture space devoted to a single species.) Secondly, maintaining strict quarantine would be critical but tough. Critical because they could well be sick (and at the very least were newborn, vulnerable, and in need of quiet and sleep); tough because babies are cute and everyone would want to love all over them. Finally, Keefe and I knew from experience how few people would step up to volunteer for the middle-of-the-night, bottle-feeding shifts. Newborns need to eat every few hours. While the idea of feeding newborn animals in the middle of the night sounds romantic, trust me—the bloom wears off quickly when one does a few 2 a.m. shifts in a row.

Still, the situation sounded grave, and we had the resources we needed to support these tiny lives, so on July 28, 2012, Russ and Leah drove to Accord to pick them up, while the rest of us prepared their quarantine space, checked our medical supplies, and became giddy with excitement.

The arrival of new animals is always a celebration. Whether we're welcoming a single blind duck, ten starving horses, three hundred chickens, or an assortment of animals seized from a hoarder, there's a palpable energy on the farm in the moments before the animals arrive. No matter how many or the shape that they're in, they serve as reminders: of why we're here, of how much we love our job, of how much that job matters.

There's a special joy in rescuing babies, of course, so when Russ and Leah pulled up to the side of the barn, I thought my heart would burst. Quickly, I glanced around at the other humans, all of whom were grinning ear to ear.

"Boys!" I shouted to them as the truck stopped.

"Hello, little ones!" Keefe said, with slightly more reserve than I. Russ encouraged us to keep our voices down, saying, "Two of these little ones are in bad shape, guys."

We instantly saw what he meant. First, the newborn calves weighed no more than twenty-five pounds each. We weren't sure if that was a normal size for Jerseys—we'd have to look up average birth weights. But we *were* sure of how sick two of them were. Not only were they as weak as rag dolls and unable to stand; they also had no swallow reflex.

We sat with them for a long while, encouraging the two who could eat to figure out how to use a bottle, discussing treatment options for the other two, and falling in love. "Wow . . ." Keefe mused at one point. "I guess I didn't realize they'd be utterly helpless." Just like any other newborn, these babies needed their mothers. Minus that option, their lives were literally in our hands.

"What do you think is going on, Jenn?" I asked. As we sat and watched, we saw that all of them had scours.

"Obviously they're dangerously dehydrated," she said. "But I'll bet more than that is going on." As we talked, one was fading before our eyes, so within a few minutes, Jenn and Russ were in the truck, a sickly calf in Russ's lap, on their way to the vet.

～

The next few weeks were exhausting for the small crew of people charged with caring for the boys. First, due to a terrible miscommunication involving a neighbor, two of the calves we had picked up were not the ones the owner wanted to surrender. She brought us the two she intended to give us, but the two who were receiving round-the-clock care had to be returned. Legally, there was nothing we could do. This was not a cruelty case, despite our belief that the animals being returned would not survive.

We also had four sick newborn calves to care for. That the calves had E. coli was obvious to our vet. We asked for a test for Giardia, a cyst-shaped organism that enters the bloodstream via the intestines and, like certain strains of E. coli, is zoonotic—it can be passed from animal to human. Our vet insisted that the test was unnecessary. We treated for E. coli.

Minimizing foot traffic in and out of their quarantine area was imperative. Our equine vet was adamant that Jenn and Keefe, the two people responsible for the health of all our animals, not handle the boys and urged us to keep the number of people who were in contact with the calves to a minimum. "You don't mess around with Giardia," she said.

So it fell to Robyn, Erin, Russ, and intern Jamie Becker to do the bulk of the "heavy lifting"—keeping their stall immaculately clean and feeding them every four hours, suiting up in full HazMat gear each time they entered the stall, bleaching their feet, the cleaning tools, and vehicle used for cleaning the stall—every single day. "It got old," Erin admits, yet she and the others did it with grace and caring . . . and soon all got sick with stomach ache, stomach cramping, and diarrhea. So did Jenn, so did Keefe. We moved the calves to our new farm just five minutes away, where effective quarantine would be much simpler. The humans recovered; the calves did not. After a full course of antibiotics, they were still sick.

We drew more blood; the vet tested for Giardia. Sure enough, the test came back positive, and the boys were put on a new regimen that included treatment with an antibiotic called Metronidizole. We moved them to a new location after three separate three-day treatments because the ground where they'd been housed was contaminated with Giardia. We'd lost precious weeks when they would have been recovering had they been treated properly. The new diagnosis meant more weeks in quarantine and all its limitations, more weeks of exhausting procedural care to ensure that the rest of CAS stayed safe. It was an anxious time.

Keeping the number of folks exposed to them to a minimum was imperative. So we thanked volunteers Donna Albright, Melissa Bamford, Betsy Messenger, and others for their help with night-time feedings, reduced the number of bottle feedings from six to four (fortunately by this time, the calves were mildly interested in hay and grain), and reduced their overnight feeders to a handful of key folks, particularly Robyn Welty and intern Jamie Becker, with Rebecca and I as backup.

~⁔

The alarm sounded at 4:30. The 5 a.m. slot on the "Calf Feeding Sign-Up Sheet" was always the last one to fill; as an early riser, I was happy to do it.

I measured out both the milk replacer and the warm water into a large mixing bucket, stirred well, and divided the formula evenly into the four half-gallon bottles. I popped on the thumb-sized nipples. Though it was a cool mid-September morning, I wore only shorts and a sleeveless shirt—one simmered when covered head to toe in an isolation suit. I pulled on the garb piece by piece—suit, booties, latex gloves—before heading to see my friends.

Big brown eyes looked up at me, blinking, as I climbed over the wall and into the stall. "Hi, sweet ones," I whispered, and instantly I was surrounded by four calves anxious for breakfast. Some staff could feed two calves simultaneously. Not me; I needed both hands to hold the half-gallon bottles. It was one calf at a time, a feat that continuously got more challenging as the calves got bigger and stronger. Seven weeks old now, they weighed at least 100 pounds; Russell, far larger than the other three, likely weighed close to 150. When he approached to be fed first, the others backed down. Well, sort of. What they really did was lick and suckle any part of Russell or me that they could get in their mouths: Russell's tail, his ears, his cheek. My gloved pinky finger. My pants. If I were kneeling, they licked my face. Sucking on *anything* was how they managed the stress of having to wait their turn.

Waiting, mind you, only lasted a very few minutes. The boys sucked hard. They drank quickly, draining the bottle dry and looking for more. I used my butt to keep one calf away as I held the bottle between my thighs and let the two others chew my hands so that that the feeding calf could eat in peace. Keefe described the process best: "It was amazing how quickly feeding went from 'Oh, let's encourage these vulnerable needy things to drink' to '*Holy crap* . . . I'm gonna get run over!' and then *boof,* there you were, smashed into the wall with cow tongues all over you."

Keefe's best feeding story was about being stopped at a sobriety checkpoint on her way home from a 2 a.m. feeding. "What are you doing out so late, ma'am?" the officer asked.

"Officer," she said, "I've just been a human pacifier, and I'm up to my elbows in cow spit."

It was clear that the antibiotics were doing their job. For weeks, while we were treating one illness but unwittingly ignoring a second one, both Calvin and Emerson's appetites were inconsistent. Emerson in particular was often lethargic and sometimes reluctant to take the bottle. But those days had long passed. The boys' energy was high, their appetites fierce, and their poop at least *looked* like normal calf poop—it was no longer soupy and yellow. We brushed them, we played with them, we sat with their heads in our laps . . . and we apologized that when we left their small stall, they couldn't come with us.

"How do you tell them apart?" people asked. They'd seen our videos of four tawny calves with no distinguishable markings. But it was simple. I've already mentioned that Russell was much bigger than the rest. Emerson is the darkest of the three—he's the color of a chestnut horse, and he has two faint white marks on his side. Calvin is the lightest in color, and Bernard is the smallest.

Beyond subtle physical differences, they are four distinctly different beings. Russell is the leader. He's the first one to approach a new situation—to walk past something scary, for example. "It's okay, guys, I've got this," his behavior suggests, and then the reassured little herd will follow. Russell loves humans and is especially good at communicating: He licks the part of his body that he wants to be scratched. We're all relieved that he's gentle! Bernard, slightly cautious, may be the brightest of the bunch. He's exceptionally aware and observant and assesses each situation before moving forward. Little Calvin is reserved and serious, often preferring solitude, sometimes enjoying his friends. Emerson? He's my favorite (though I don't tell the others). He's curious, a bit pushy, and the most affectionate little guy I've ever met. He'll give a greenhorn a run for her money, and he kisses everyone he meets.

～～♪

"The second negative fecal came back," Jenn announced when I walked into the barn.

"GET OUT!" I exclaimed. The wait had seemed endless; how we felt for the growing boys locked in a stall in an otherwise empty barn. But with this extra confirmation that the calves were healthy, they could finally return to the main farm. No more suits, no more gloves; much more importantly, no more isolation. Better yet, they would soon be ready for what we labeled "the big boy field," a pasture on high ground at the northern end of our eighty-acre main farm. With a large hilly field and loads of room to run, horses and cows as neighbors, and a view of nearly the entire sanctuary, we knew we'd soon have four very happy campers.

～～♪

When the day arrived, eight of us walked out to witness the moment. Volunteer Phyllis Kaiser led little Calvin; volunteers

Donna Albright and Sue Rich led Emerson and Bernard. Russ, naturally, led his boy Russell. Leah was there, video camera at the ready; Jenn, Robyn, and I were there, too, because, well, we wouldn't have missed it for the world.

Russ opened the gate and walked in with his "son" and name-sake Russell; the others followed quickly behind them. "It's your big day, boys," I said. Others offered words of encouragement as we slipped off their nylon halters.

Cattle aren't generally the most energetic of animals. They spend a good portion of their day lying down and chewing their cud and another good portion grazing. Unlike many horses, who run for the sheer joy of being alive, cows run if they're fleeing a predator or other threat. If they're really hungry, they'll run toward dinner. Otherwise, they don't run—unless they're excited.

Barely before we could remove their halters, the boys had sized up their roomy new digs and began to run. They ran, and they ran, and they ran, and then they ran and they ran. And then they ran some more. They ran along the full length of the pasture. They circled back across the middle of the field, barreling straight at us, swerving to our left or right just in the nick of time. They ran in circles around the barn. They ran up the hill, down the hill, occasionally hopping or bucking or kicking out in glee . . . but mostly just running, because for the first time in their lives, they could.

We humans stood for a few minutes, enjoying the reward of four months of hard work. Hand-walked and socialized every day for the first months of their lives, these calves will have important jobs as ambassadors for their species.

As we exited, Emerson stared through the fence at Dozer, a full-grown Jersey standing just feet from him in a neighboring pasture. For a moment, Calvin nosed one of the rubber balls we had given them as toys, but it was far less interesting than the big pile of hay into which Bernard and Russell's faces were now buried. He walked over to join his friends.

"We did a good thing, guys," I said to the group as we left the pasture. On the day that these boys' lives were really beginning, I wondered if my human buddies were thinking about what I was: that on this same day, the lives of so many others just like our four special boys were ending. As Calvin, Emerson, Bernard, and Russell experienced their first taste of freedom, approximately 2,700 calves their age or far younger (most veal is "bob veal," the meat of newborns) were ending a life of isolation, deprivation, and poor health by hanging upside down by a single leg from a chain, their throats slit as life bled out of them.

~~~

I recently discovered an industry website called vealfarming.com. I poked around for a few minutes, and found a Q&A section that defended standard industry practices. Sections of it are included below, and while my response is on the right, the industry's justification of its practices probably speaks more loudly and clearly than anything that I could say.

What's also noteworthy is that the veal calves who are depicted on this site are a small percentage of the whole. Most calves, perhaps as many as 90 percent according to Harold Brown, former dairy farmer and president of Farm Kind, are immediately taken from their mothers and shipped to auction. They are dead when just a few days old.

Does this sit okay with you?

Sweet, sweet Russell.

Erin loads the truck for morning feed.

Caleb (L) and Patty (R). Caleb was rescued by a concerned neighbor; Patty was one of six animals CAS was delighted to rescue when Catskill Game Farm closed in 2006.

Lorraine relaxes in a flower bed.

Claude the pig greets the day.

Ozzi with a young guest.

Allen and Dozer. Like Dozer, many cows are extremely affectionate.

A few members of the Underfoot Family, our ever-changing group of animals free to roam the entire farm, graze in front of the horse paddock. Underfoot Family from L to R: Casey, Rambo, Hannah, and Aries.

Hope (L) and Echo (R) play gently. These two mares are among fifteen rescued from an irresponsible Saratoga breeder. Our farrier described the horses, who arrived thin and with wretched hooves and mangled manes, as "semi-feral."

Keefe and Petita share a snuggle.

A view of equine alley, the southernmost section of CAS

Former Animal Care Director Abbie Rogers with Betty, a few days after her birth

Betty's brother Nash. In "Just Another Day at CAS," I tell the story of their mom's rescue.

Robyn helps her goat buddies reach a treat.

"All of this food is cruelty-free!"
During our events, these signs are posted at
our food tables.

Rambo and Hannah on early-morning patrol

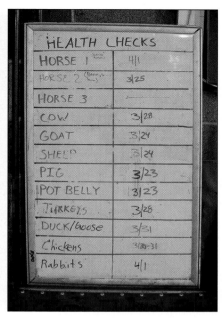

HEALTH CHECKS

HORSE 1	4/1
HORSE 2 (Barry)	3/25
HORSE 3	—
COW	3/28
GOAT	3/24
SHEEP	3/24
PIG	3/23
POT BELLY	3/23
TURKEYS	3/28
DUCK/GOOSE	3/31
Chickens	3/30-31
Rabbits	4/1

Buddy, who is blind, is featured on the jacket of my first book. He arrived at CAS with a dangerous panic disorder; his surrenderer did not tell us he'd been hit by a car.

Erasable white boards help all staff stay abreast of various daily and weekly routines.

Troy brings little Ozzi in for the night.

Jangles: now THAT'S a happy pig!

Rambo and his pal Barbie

Noah, a special friend

Gentle PeeWee, removed from a hoarder in 2004. Most of the woman's goats had already died of starvation.

Alex and tiny Lux. Our rabbits have come from a variety of distressing situations: dumpsters, fur farms shut down for cruelty, etc.

My pal Franklin, back at CAS after his summer at Animal Camp

"Who says I can't hang with the big guys?" Free to roam the entire farm, Malachi chooses the horses as his friends.

Sweet Norma Jean, rescued the day before Thanksgiving

Gentle Hank, immobilized by industry-induced obesity, shares a tender moment with Animal Care Director Jenn Mackey.

Special-needs goat Atlas cuddles with his beloved friend Henrietta.

Fritz and Phyllis share a laugh.

Patrick and Phineas: *Isn't life grand?!*

Volunteer Julie Buono and I visit with Casey the horse, who lived in a junkyard and arrived at CAS thin, wormy, and covered in ticks.

Russell and Russell: to see them together is to see love in action.

How I love him.

Camp Kindness kids enjoy pickling with Judi.

Little ones enjoy a peaceful evening.

Best friends Gertie and Lola.

Amos, rescued from the notorious Catskill Game Farm, grazes on a cool spring evening.

Deeply affectionate Zeke is one of a large flock of sheep seized from an illegal backyard slaughter operation.

Photographer Jill Meyers with her pal, Arthur.

From vealfarming.com	My response
Veal farmers raise calves in barns, where they are protected from weather, predators, and disease.	Yes, they're protected from predators. But they're also deprived of their mother's love, their mother's milk, sunshine, and the ability to run, play, and discover the world like any newborn calf longs to do. *Protected from disease?* By depriving veal calves of their mother's milk, fresh air, grass and hay, exercise, proper nutrition, and veterinary care, veal factories are breeding grounds for infectious disease, as we learned firsthand with our four calves. Everything about this model is wrong. Newborn calves should be raised by their mothers—*period.*
Calves in individual housing can comfortably lie down in a natural position, stand up, groom themselves, and interact with neighbors.	It's a pretty big stretch to call veal crates or huts comfortable. How is it acceptable to give any living thing only enough room to stand up and lie down? Imagine if you were forced to live your entire life in a box just inches wider and longer than your body. How long would it take before you broke?

From vealfarming.com	My response
In individual pens, calves can stretch without fear of other calves stepping on them.	I'm sorry, but *give me a break.* Little ones are clumsy. Sometimes they step on each other. *So what?* Chained or locked in tiny crates or huts, veal calves can't walk, stretch their legs, play with friends, run, or lick and chew on each other as Russell and friends have taught us they love to do. And again: *They don't have Mom.*
Many farmers have adopted group-housing methods to provide the same level of individual care as in traditional barns.	Vealfarming.com claims that 30 percent of growers have adopted group-housing methods. I urge you to look at their website and check out the video of their model "group housing" farm. What I saw was pretty wretched.
Individual stalls maximize the quality of individual care farmers and veterinarians can give the calves. Also, most importantly, minimizing calf-to-calf contact is the best prevention against disease.	What quality of individual care? Being nurtured by their mothers as nature intended? No. Proper food? No. Proper shelter? No. Grooming? No. Warm, soft bedding in which to lay? Time to explore their environment and interact with other cows? No. Medical care to counter the impact of the hideous environment and diet? No. *What quality of individual care do they refer to?*

From vealfarming.com	My response
Veal calves receive diets designed to provide all of the forty essential nutrients they need, including important amino acids, carbohydrates, fats, minerals, and vitamins.	Veal calves receive diets designed to force rapid growth and keep their flesh lean, pale, and tender. No aspect of their diet is for their health. They are denied their mother's milk, denied all solid food, and denied drinking water. A proper diet would not cause scours, would not cause anemia, and would not produce *white flesh*.
Both male and female offspring of dairy cows are normally removed from cows soon after giving birth. This separation allows dairy cows to return to the herd and produce milk for human consumption. While calves are not with the dairy cow following birth, they still receive her colostrum, or first milk, within twenty-four hours.	If someone took your child immediately after birth, what would you do? We take newborns from exceptionally maternal animals and immediately hook the bereft mother up to a machine that takes the milk she made for her calf. And what must that calf feel? To experience firsthand the helplessness of Calvin, Emerson, Russell, and Bernard, and to know that two would have died without round-the-clock emergency medical care, was a real eye-opener.

Since 1986, the reduction in veal consumption by close to 75 percent (from a high of 3.4 million calves slaughtered that year to the current rate of under one million) highlights the need for effective educational campaigns about *all* animal agriculture industries. Beginning that year, the Humane Farming Association began an unrelenting campaign against the veal industry that included television advertising, full page ads in national print media, and more. My guess is that if you're still eating animals, you don't eat veal because of the success of HFA's campaign. While I'm not sure why they focused on formula-raised veal as opposed to "bob veal," the flesh from calves slaughtered almost immediately after birth (since bob veal comprises most of our consumption), perhaps it doesn't matter. What matters is the success of their campaign. What matters is that we heard them and veal consumption plummeted.

I sometimes wonder if death comes as a relief to animals used to feed humans. Whether calves, piglets, chickens, turkeys, ducks, geese, sheep, or other food animal, food animals' lives from beginning to end are so horrific that perhaps death is a welcomed escape. I don't know. What I do know is that among the approximately 10 percent of veal calves given a few months of life, those lives are less wretched than the lives of chicks grown in warehouses and slaughtered for meat, laying hens stuffed into battery cages then slaughtered after a few years when their bruised, spent bodies are used up, and so on. The way I see it, what we subject all of them to speaks of an industry utterly devoid of conscience and a larger culture too willing to turn a blind eye to the consequences of our dietary choices.

There *is no campaign* against these industries that even comes close to the size of HFA's anti-veal campaign. But the moment you draw your own line in the sand is the moment one hundred animals per year (assuming you have a traditional diet) are spared this unspeakable suffering. Please: Consider being your own campaign for the animals.

It's a Hot Day Here

I don't see a single leaf waving. It's probably ninety-five degrees today; the air is thick and still as death. But I don't turn on my air conditioner, for quite simply, it uses too much electricity. Global warming. The earth is pleading for our help, so I'm making concessions where I can. Like turning on the hot water heater only thirty minutes before my shower. Like pooling my errands so I drive less. Like deriving the Sanctuary's energy from a newly-installed solar system. And yes, like going without air conditioning when I want it. Being vegan, of course, is far and away the biggest thing I do—the biggest thing any of us can do. Lessen the foot-print. I'm trying.

The animals, after all, don't have air conditioning. They have shade, sure, but on days like this, they'd love to be standing right in front of that machine that magically takes hot air and turns it frigid. Today, we all sweat. I sweat at my computer, and Julie bakes in her office, refusing, bless her conscientious heart, to turn on her air conditioner. The barn crew all look like they've been playing under a sprinkler—and no doubt wish they had.

Among the animals, the pigs fare best. They loll in the mud baths we make for them under the willows, find the shade trees and dig up the cool earth, and grunt with glee when, as they leave their pasture to come to the barn for nighttime, Keefe sprays them with cool water from the hose.

The sheep, recently shorn but nonetheless trapped inside their dense coats, pant heavily. The horses and cows find the shade and stay still, but two elderly cows are nonetheless dangerously close to heatstroke. Yet it's the smallest of all, the poor broilers, who struggle the most. They've got the coolest spot on the farm—their house sits under a canopy of enormous willow trees, and their yard is literally a foot from the pond. Still, as the thermometer inches toward a hundred degrees, every single broiler is holding her wings away from her body and panting heavily. We carry them, one by one, into the barn, and stand them in shallow bowls of cool water, hoping their body temperatures will drop.

Chickens, you see, aren't supposed to weigh fifteen pounds, nor are they supposed to have breasts so large that at any moment they look like they might pitch violently forward and break their necks. But the poultry industry began manipulating chicken genes to create big meaty birds with huge breasts right after World War II. Sixty-five years later, every aspect of their production, housing, and feeding is designed to grow the largest possible bird in the least possible time: between thirty-nine and forty-two days, to be exact.

Agribusiness has created true "Frankenbirds" to grow at freakishly fast rates to maximize profits for the producer. With their systems so horrifically stressed, however, broilers have such high death rates that agribusiness itself has created the term "flip-over syndrome" to describe the young lying on their backs, feet pointed skyward, dead because their systems can't take the rapid growth. A number of our own birds have died in this way.

Still we humans continue to eat them. Tortured baby birds, killed at *seven weeks old*. They still have the "peep" of chicks when we send them to their deaths. Caged pigs, terrorized cows. Under agribusiness, the tiniest concessions to their well-being are long gone. The animals are commodities, period. It doesn't matter that they suffer mightily. It doesn't matter that they are so very much like us, or that pain is pain and suffering is suffering whether it is inflicted on a human or a dog or a chicken. The poultry industry doesn't give a rat's ass about the well-being of the birds.

So just like the 300-pound human is having a harder time on this hot humid day than I am, so are our chickens. They're desperately gasping for air.

The Slaughter Truck

Good farm animal sanctuaries are the antithesis of agri-business. This is true regardless of the species one considers. A sanctuary's work is to help animals thrive; factory farming's work is to ensure that corporations thrive. Let's do a little comparison and contrast of the lives of pigs.

At Catskill Animal Sanctuary, every day is Pigs' Day. Our pigs sleep in heated barns on cold winter nights and wallow in mud on hot summer days. Once in a blue moon, if it's truly sweltering, we open the gate to the pond and watch as 1,000-pound pink packages glide through the water as smoothly as swans. Our pigs are grouped based on friendship, for pigs who don't like each other can do serious damage in the blink of an eye. We clean pigs' ears and eyes, trim hooves and tusks, exfoliate dry skin, include vitamins and produce in their diets. When they look a little too pink in mid-summer, we slather them with sun block. We go to lengths most might consider extreme to ensure not only that their basic needs are met, but that their emotional needs—their varying needs for friendship and connection—are met, as well. A mom wants her child to be well-fed, safe, secure, and happy. Many of us want these things for our companion animals. Animal sanctuaries want them for all our animal friends, because we want them to thrive. This desire is at the heart of all that we do.

Agribusiness, which grows 99 percent of the meat consumed in the United States, has the opposite desire. Its goal is for the corporations to thrive. No matter the animal, every aspect of the system is designed for maximum efficiency. The more quickly a company can grow its product, the fewer resources required to do so, the greater the number of animals packed into transport trucks, the quicker the slaughter line moves … then the greater the corporations' profits. It's capitalism at its finest.

But we're talking about living beings here, folks. In the case of pigs, pork producers know who they're dealing with. Pigs are famously bright, willful, and strong enough to drag a man down the road as if he were a field mouse. Every single aspect of turning strong, *threateningly intelligent* animals into bacon and sausage has been tweaked and honed to maximize profit and to minimize physical risk to the humans who handle them. Never mind the pigs' *monumental* suffering. It isn't part of the equation.

Transportation to the slaughterhouse is one horrible piece of a piglets' nightmarish life. Piglet? *yes.* Turkeys, chickens, and veal calves are only weeks old when they're slaughtered. Piglets are slaughtered at six months. They are beaten with metal pipes, bats, or two-by-fours, or shocked into compliance with hand-held electric prods that are sometimes jammed into their rectums. Terrified for their lives, they rush onto the trucks that carry them to their deaths, then are packed like Vienna sausages for the trip. *They all scream.* More than 100 million pigs are killed annually for

Americans to consume, but the pork industry also reports that a million more die each year during transport.

Does it surprise you that livestock transportation is a largely unregulated industry? It shouldn't. It fits the model: do it fast, do it cheap. A hauler can carry as many animals as producers can beat, throw, or shock onto a truck. Some pigs fall and are crushed or suffocated; others die of heart attacks. Death from exposure is equally common. During sweltering summer months, many die from heat exhaustion; during winter, pigs standing around the perimeter sometimes freeze to death, stuck to the sides of the trucks. Others, still alive, are frozen to the truck and unable to move. A 2006 industry report quotes a slaughterhouse worker describing how workers "go in there with wires or knives and just cut the hogs loose. The skin pulls right off."

〜〜〇

I've read about livestock transportation for years. But reading about it and witnessing it are two different things.

"Don't take Route 81," my dad cautioned on the phone one night when we discussed my route home from Nashville, Tennessee, which was the final stop in a six-city book tour and a great excuse to see my ninety-three-year-old grandmother and some other beloved relatives. "It's a truck route. Scary as shit to be boxed in by three tractor trailers going seventy-five miles per hour."

I considered Dad's advice. I didn't relish the idea of driving nearly six hundred miles, much of it mountainous, surrounded by trucks. But the other option, driving *way* east via I-64 then heading up I-95, would add close to a hundred miles to the trip. So at 6 a.m., I said a teary farewell to my aunt Beverly Ann, her husband,

Frank, and mutts Bailey, Sammy, and Levi, and then pointed the car toward Route 81.

Around 1 p.m., with the sun high in the sky, the day warmed. I opened the windows . . . and that's when I smelled it—a slaughter truck, climbing the hill in the slow lane as I approached it on the left.

I don't often travel long distances via interstate highways, so I rarely encounter these deathmobiles. I've seen chicken transport trucks jammed so tightly with crates of chickens that many have suffocated by the time the animals arrive at the slaughterhouse for their barbarous deaths. Long before I began the work of trying to raise awareness of these delightful beings' sentience and the depth of their suffering, I wept when I passed the trucks. Aluminum boxes on a flatbed, rows of oval holes cut into their sides for ventilation. But that's all—that's the single accommodation for the animals, and that's done only so they won't die en route to the place that will slit their throats, plunge them in boiling water, rip out their feathers, neatly slice off heads and feet, and clean and package what remains of their battered bodies for humans who would be far healthier if they did not consume their poisoned flesh laced with pesticides, antibiotics, and growth hormones.

Today, I did more than weep. You see, I know who these animals are now.

In my first book, I recalled the life and lessons of one chicken named Paulie. Paulie was the barn peacemaker, a frequent passenger in my car (I usually insisted that he ride shotgun, though my lap was always his preferred seat), a good friend to my dog Murphy, and our regular companion at lunch. There have been other chickens, too—birds so full of quirky personality and a desire to communicate that one swears they really have vocabulary *if only we could understand it.*

Today's truck was stuffed with pigs. Stuffed so tightly that what I looked at through the ovals was just a solid mass of pink. No doubt snouts were jammed into rectums and sharp hooves into tender skin; no doubt that beneath the solid mass of pink, others had already succumbed.

So as I passed this truck carrying animals I know to be uncannily "human," one pig caught my eye. He looked at me through the oval hole; the look shared more than words ever could. In that moment, he was every animal ever grown to feed humans, and in this helpless, hopeless moment, he was asking a simple question: *Why?*

A wail emerged from my body. Not just tears. An uncontrollable wail—I couldn't stop it—coming from a deeper part of me than tears ever have, and an apology to that pig, and to all animals on behalf of my species.

I will return to CAS, where I will hug my pink pals Franklin and Policeman and Claude, and they will love me right back, with smiles and happy grunts and snouts rubbed into arms and cheeks, so that within a moment, I'll be happily as muddy as they. And I will wonder about my good fortune to be born a human and not any other kind of animal.

The power to change this, folks, is in our hands.

On Love, Loss, and Hot Dogs

Last Sunday was fun. I was one of three judges at Go Dog Go, a fundraiser held by Pause Dog Boutique to support dog rescue efforts.

It was fun because I was in the company of Bruce Littlefield, "arbiter of fun" and author of the best-selling book *The Bedtime Book for Dogs*, and Pia Salk, a psychologist who specializes in the human–animal bond and is the spokesperson for Adopt-A-Pet. Honestly, judging stool samples would have been fun with these two, but when you add the fact that the divisions were "Bad Hair Day," "I Look Like My Mama," and so on, and that dog kisses were available on demand, well, the day was custom-made for me.

Go Dog Go would have been my idea of a good time anytime, but last Sunday, it was especially meaningful after a tough week of saying goodbye to my beloved Babe, a seventeen-year-old steer who touched thousands of lives at Catskill Animal Sanctuary.

Babe, all 2,000-ish pounds of him, arrived at CAS in 2003. We were a fledgling organization . . . and looked it. We had just purchased our property—a forlorn, forgotten farm that needed rescuing every bit as much as the animals who would soon arrive—and were busy hauling away piles of tires, toilets, and rusted vehicles when big Babe barreled down the drive.

Babe's human friend had rescued him at auction, wanting to save one life. She'd bottle fed him, kept him in her back yard until he simply grew too big, then boarded him, rather ironically, at a beef farm. When she divorced, she could no longer afford his board, and the farmer gave her a month to find him a home. Fortunately for Babe, and fortunately for us, his human found Catskill Animal Sanctuary.

I am flooded with memories of the placid being with soft eyes and a tongue that reached out expertly to grab one carrot, then another, then another until a five-pound bag was gobbled up. Among my favorites:

- I'm not sure why, but for a short while, Babe was a free-ranger. Only Babe, theoretically free to roam the entire property, never went *anywhere*. Babe instead stood in the middle of the barn aisle, motionless as Peepers the duck patrolled the barn by waddling frantically back and forth, under Babe's belly. "Quack-quack!" Peepers said, rushing under the black giant, who stood unfazed, eyeing the feed room. Surrounded by free-ranging ducks, chickens, goats, sheep, and a growing human fan club, Babe seemed supremely at peace.

- A bunch of us were cleaning the cow field. I stood atop the tractor bucket to bellow instructions to volunteers who were spread throughout the large pasture. Babe wandered up to say hello, but got so close that when he turned his head to flick a fly away, he sent me sailing through the air. I landed with a thud, gathered my breath, then laughed hysterically. Babe startled, but then walked over and licked my head.

- 6:30 a.m. Two police cars pulled down our driveway. "Are you missing somebody?" a young cop asked. I remember the wry smile on his face; somehow I knew that this was code for "Your cows are in town." In town they were, all right . . . Babe had led his herd through the woods, down our local rural lane, and onto very busy Rt. 9W . . . and then just stood there. The escapade, and our return walk home (I put a bright green draft horse halter on Babe and led him down the road; the others followed) was on the evening news.

Perhaps one of my most favorite memories of Babe occurred several summers ago, when the renowned Omega Institute invited us to provide animals for their animal communication workshop. We took our best ambassadors. Whether we learned how to "communicate" with them (I believe all of us are communicating with each other all the time but am ambivalent about how "animal communication" is often taught and have certainly questioned some of the feedback we've received from animal communicators), I wanted our animal friends to open the hearts of those whose worlds had only included companion animals.

I didn't expect that Babe would want to join us. Yet when I opened the trailer door, he evidently did, because he stepped right in and cozied up next to Chester the horse. That he had absolute trust was remarkable enough. Is it a stretch to say that on some level he knew we needed him? I don't know.

Throughout the weekend, the communicator guided us through various relaxation techniques, after which we would open ourselves to the messages our friends wished to share. While I was unimpressed by the workshop, I was blown away by our animals' comfort level and their obvious connection to their CAS people (we'd been given six slots). I learned a great deal from them.

The instructor had decided that Babe would be the final animal with whom we'd practice. And this is exactly what happened:

We had just finished "speaking with" Chester, the funny old Appaloosa, and in the meadow where we'd gathered, Chester put his head down to graze. "Let's turn our attention now to this big guy," the teacher said, gesturing toward Babe. As she suggested that we form a circle around him, Babe walked right into the middle. Then, as she'd done with each of the animals, she invited us to close our eyes. As she spoke softly, my eyes stayed open, and I watched a huge black head droop.

"Relax," she whispered, and he did. Babe's giant head dropped lower, and lower still, until right before his knees were about to buckle, he folded his legs and lay down.

And that's when I received a clear and plaintive and powerful message. "We don't want to be hamburger," is one part I remember, along with "I hope I'm doing a good job. Thank you for choosing me."

To this day, I'm not sure what to make of that moment.

I don't recall what anyone said about what Babe "communicated" to them. But I do recall many of them weeping, one woman sitting in front of him and cradling his massive head as he licked her face, and another woman draping her body over Babe's back (after she saw me do it). And I remember the thought that struck me like a lightning bolt: "These people will never eat meat again."

❧

I suppose the organizers of Go Dog Go thought it would be clever to sell "hot dogs" (and nothing more) at a "dog show." But the "hot dogs" were not hot (as in cooked) "dogs" (as in canine) at all. They were cooked cows. I'm fairly certain that my Babe, and every other cow or pig who has ever lived, if given the "Bad Hair Day" or "Become a Hot Dog" choice, would have chosen the former.

I stood in line behind half a dozen animal lovers, all of whom ordered cooked cows, and none of whom noted the contradiction in eating animals while attending a fundraiser to benefit animals.

"I'll have the veggie burrito," I said to the vendor, who had put them on her menu at our request.

"I have a pet pig," she said as we chatted briefly about my special request. "He's really something."

I've been invited to be a judge again at this spring's Go Dog Go. I told them I'd be delighted to be included if CAS could provide the food. I've not yet heard back from them.

Rambo's Last Day

In *Where the Blind Horse Sings*, readers met a sheep named Rambo, who arrived at our fledgling sanctuary in 2001 filled "with testosterone and rage." With the help of the New York State Humane Association and the Ulster County SPCA, we rescued Rambo and sixteen other animals from the derelict property of a local hoarder, where they were confined in a single, filthy stall and fed an infrequent diet of stale bagels. Rambo had not been too pleased with this arrangement. For more than a year, he made his distrust of all humans clear by rushing, head down, toward anyone who came near him. If you'll look at his photo in the center pages, you'll understand how dangerous it was to be near him.

With love and time, however, Rambo's rage disappeared, replaced by a driving need to protect all who lived and worked at CAS. Think about it: It's in the DNA of rams to protect their flock. What fascinated and moved us for so many years was Rambo's choice to include *all who lived and worked at CAS* in his flock. He didn't simply look out for the additional free-ranging sheep. That choice would have been natural and endearing, but unremarkable. Instead, Rambo kept a watchful eye on all CAS beings. *That* choice, borne out in extraordinary acts of compassion and courage, inspired admiration and awe in those humans lucky enough to know him. What's more, his ability to communicate with us was so striking that it demanded a change in our perception of the differences between human and non-human animals. At least that's what it did for me. His influential lessons have shaped many aspects of our rescue operation along with my presentations on animals at places from kindergartens to colleges to national conferences. Rambo's life mattered.

Thousands of you knew and loved this remarkable beast. You marveled at his metamorphosis, as depicted in *Blind Horse*, and at his continued evolution documented in the first printing of *Animal Camp*. You visited him—some of you several times a year! Those of you who've been friends from the early years were lucky enough to observe his evolution from a violent animal to the one who walked into the middle of every tour, rested his head on someone's thigh, and pawed that person's foot . . . over and over again.

"He wants you to scratch his butt," we had to explain to the uncertain guest. And of course there's a story in this . . .

Rambo was standing in the middle of the barn aisle during a nightly barn check five or six years ago. His head was turned as he scratched his rump with the tip of his massive horn. The itch must have been a fierce one—he was scratching hard.

"I'll scratch your butt, Rambo," I said. Little did I know that this single moment would set up a pattern that lasted until the end of his life.

I stood behind Rambo, scratching the top, back, and sides of his derriere. I scratched with my nails for a bit, then kneaded his itchy skin with my fingers and watched his head drop and eyes droop in relaxed enjoyment.

"Yeah, Rambo . . . feels good, doesn't it?" I said to him before heading to the kitchen—before *attempting* to head to the kitchen—to retrieve treats to distribute to the animals.

You see, Rambo was blocking my path. I *couldn't* go to the kitchen, because he wouldn't let me. "Excuse me, Rambo, I have to check on the animals," I said as I tried to move around him. Of course I knew what he wanted, but to challenge his ability to communicate, I feigned ignorance.

Every single time I took a step, he did, too, so that he was continually sideways in front of me, physically blocking me from going forward.

Oh my. What to do next? Granting his request for more butt scratches wouldn't answer my question. If I kept pretending not to know what he was asking for, what more might he do to communicate what he wanted? Here it was: another teachable moment for both of us. I wasn't about to overlook the opportunity. So, I just stood there.

When acting as a barrier didn't yield what he wanted, Rambo faced me and pressed his forehead into my thigh. Okay, pretty cute.

"I love you, Rambo!" I said half a dozen times, because it was killing me not to give him what he wanted.

I imagined him thinking one of three things:

1. *Could I be any clearer here?*
2. Patience is a virtue. Patience is a virtue. Patience is a virtue.
3. Good *lord*, I thought she was sharper than this.

Still I stood there, frozen and giggling. In a minute, Rambo turned sideways again, and this time, he stepped on my foot. Hard. Oh my *word*. My hands remained at my side.

When Rambo pawed my foot over and over a few seconds later, I gave in. I simply couldn't resist any longer. Besides, I figured the only thing left in his repertoire would be to flatten me to the ground or more likely to give up and feel defeated (because he wouldn't be willing to knock me down). So I knelt beside him and gave him a good massage, enjoying his pleasure and the sensation of his pliable skin beneath loose wool. I marveled, once again, at this angry man turned healer, savior, hero, teacher—and I laughed my fanny off, too.

For the next few weeks, I made a point to be in the barn quite a bit, because from that night on, Rambo marched up to *anyone* whom he sensed would grant his wish and pawed the person's foot.

"What are you doing?" folks asked him. Either that or, "What are you telling me, Rambo?"

I explained and demonstrated the technique, and Rambo got his wish. In fact, Rambo got his wish many, many times a day for the rest of his life. *Thousands* of massages. It wasn't long before he figured out that all he needed to do was paw the ground wherever he happened to be and someone—or several someones—would go to him and, yes, scratch his butt. He trained us well.

So here we were in May of 2012. Rambo was at least sixteen—that's an old, *old* sheep. How lucky we'd been to have so much time with him, especially since we thought we were going to lose him two years earlier, when his appetite was inconsistent, his arthritis severe, and his limp pronounced. But a combination of acupuncture (which he tolerated as long as I was with him) and very occasional cortisone injections in his knees gave Rambo a new lease on life, and gave us two more years of watching him age with quiet grace.

By spring of 2012, however, these interventions were no longer effective, and there were only three alternatives:

1. Put him on a narcotic like Tramadol, which was out of the question.
2. Let him live with increasing pain and decreasing mobility, and risk the likelihood of a grave, even life-threatening accident, and the very real possibility of him dying alone and in pain, which was also out of the question.
3. Swallow hard, thank the heavens for eleven years with this noble animal, and let him go. Yes. It was time.

"I need to prepare you for something," Keefe had said the previous day, taking my hand in hers. Keefe is not a hand holder. "It's probably going to be soon. Not tomorrow, but it could be as soon as a week. We're watching him very closely."

"I know," I said through my tears.

But that very night, as I walked into the barn, I saw Rambo stand up from his bed to pee. Unable to put weight on his left front leg, he struggled up on his right. But it buckled, and Rambo fell to the ground. He stood up again, unassisted, but I'd seen what I needed to see. We scheduled his euthanasia for the following day.

That night, I sat with Rambo in his deep bed of straw until well after midnight. Though they didn't last for hours, we'd had quiet moments like this at least two dozen times in the past year, and more recently, as I knew his time with us was drawing to an end. Sometimes, I massaged his wooly cheeks and the space between his horns. Sometimes I sang to him. Sometimes we sat in silence. And sometimes, as if I were reading to him, I told him the stories from his life.

Where the Blind Horse Sings shared some of the remarkable moments from Rambo's eleven years with us: The morning he helped me round up three llamas who had broken out of the barn and were on their way to Nashville. His saving of Charlie the pig, and the cold November night when he explained to me that we'd left two turkeys outside in the bitter weather. Plenty of others didn't make the cut; for instance, the day when Petunia the pig was adopted by a delightful family, and he followed the trailer up the driveway, *baahhing* desperately.

When the trailer rounded the corner, he charged down the hill straight to me. "*BBAAAAAAHHH!!*" he wailed, looking right up at me. It dawned on me that this might have been the first time he'd seen an animal *leave* CAS. That moment conflicted with his current world view: sad animals came to CAS to get happy. Happy animals were not supposed to leave CAS. Rambo was inconsolable; he followed me to this house, screaming. "THEY TOOK PETUNIA," he said. Or perhaps even "*WHY DID YOU LET THEM TAKE PETUNIA?*" Whatever the message was, he was desperate, as the leader of our flock, to communicate it to me. It seemed he felt both that she was being kidnapped and that it was our shared responsibility to prevent it. I recounted this memory as we sat in his bed.

I recalled another night when I ran into the barn to check on a sick old horse named Casey. I was on my way to speak at a local college and didn't want to be late. Apparently, though, I was going to be. Rambo, his breathing elevated, was standing in front of my car, physically blocking me from leaving. *Aaah . . . of course.* In Rambo's world, we humans run when things are wrong: the cows

151

are out; a horse is down. Rambo had already experienced watching me sprint across a field to retrieve three llamas who'd escaped from the barn—and sure enough, he came to the rescue. So once again, Rambo logically assumed something was gravely wrong. I watched as he turned in a full circle, one quick but deliberate step at a time. He scanned his surroundings, trying to determine where the danger was, while preparing to protect his flock.

"Do you remember that night, Rambo?" I asked. "You were so brave . . . you wanted to make sure we were all safe."

Eventually, the stories were done, and my eyes were heavy with sleep. It was time to go to bed. "Thank you, Rambo," I managed to whisper as my tears fell freely. "It has been an honor to be your friend."

I sat rubbing his forehead for a moment more, then leaned forward to kiss him. "We've been a hell of a team" were the last words I spoke. I hope he felt them.

The morning of Rambo's final day, Keefe and Russ found a young horse named Timothy dead in his pasture. It was the Saturday of Memorial Day weekend.

Several of us were busy preparing the Homestead, Catskill Animal Sanctuary's beautiful old inn, for its first guests. Caleb had done remarkable work over the last months transforming the sad building into a gem we could be proud of. Still, the last-minute "punch list" was a long one, and here I was, washing windows, arranging furniture, checking in with Chef Linda about the details of our vegan breakfast, mere hours from saying goodbye to the being who has had a greater influence on me than any being other than my father. Just then, Keefe walked onto the porch, tears streaming down a grief-stricken face.

"Timothy's dead," she said.

I exhaled more breath than I realized I had.

Timothy was a beloved and gentle horse who'd been rescued six months earlier from a Saratoga County hoarder. We'd accepted fourteen horses and fostered out six, leaving us responsible for eight. While they were all a mess, Timothy was by far the worst—skeletal, and too weak to step up onto the trailer the day of their rescue. We loved them all, but he was our darling: gentle, earnest, willing, and affectionate.

The truth of the matter is that, in sanctuary work, one gets good at saying goodbye. We've experienced scores of deaths in eleven years. How difficult each loss is for each of us obviously varies, depending on our own relationship with the animal, how long the animal has been with us, whether we've had adequate time to wrap our heads and hearts around his impending death, whether we've been able to say a proper goodbye. Timothy's death tore Keefe and Robyn and Russ apart. For me, the toughest one of all was a few hours away.

Mark Rosenberg, a lovely veterinarian whose Hurley Veterinary Hospital practice cares for all our ruminants, arrived at the end of the day.

"He's here," my friend David said quietly as he entered the house. I had called David earlier in the day to share the news. He came to support me and to witness Rambo's passing.

It was time. I bent over, placed my hands on my knees, exhaled sharply. "Okay," I said to him. "Let's go."

Oh, Rambo.

The barn was quiet. It always is, prior to a euthanasia. While we've never discussed this, the hush that falls over us in these moments is profoundly beautiful. To me, it feels like we're honoring the animal we're about to lose, but just as importantly honoring each other's need to work through powerful feelings of loss and grief on our own.

Robyn, Jenn, Russ, Alex, and Keefe were there, along with beloved volunteers Donna, Siobhan, and Julie. Casey, the old paint pony who has been a fixture here since 2004, looked out at his human friends. Rambo was in the only spot I'd seen him in for the last several days—in his deep bed of straw, legs tucked under him.

I sat down with him, and placed a copy of *Where the Blind Horse Sings* next to him. I had a Rambo cap—a baseball cap with Rambo hand-painted on the front—pulled on backwards to keep my hair out of my eyes. I took it off and placed it next to him as well.

"It's your turn, sweet one," I whispered to the animal whose influence and friendship throughout the years defies easy description. "Are you ready?"

A few others came over to say goodbye. Keefe whispered softly in his ear, "Goodbye, boy. It's been a privilege."

I sat, Rambo's head on my thigh, whispering "thank you" and "I love you." Mark began the two-step procedure we've seen so many times: first the tranquilizer to put the animal into a deep sleep, followed by the drug that stops the heart.

For a few quick seconds, instead of getting sleepy, Rambo became anxious. Did he know what was happening? *Was he trying to tell us he wasn't ready?* I can't answer these questions; to this day, they sometimes haunt me. Within a moment, though, he settled; his body relaxed, his eyes got heavy, and he was asleep. A few peaceful minutes later, surrounded by a few of the legions who loved him, Rambo was gone. No drama, no hysteria . . . just a lot of good people sending him off with all the love we could muster.

Several of us lifted Rambo's body onto a blanket, and David and I drove him to Hurley Veterinary Hospital, where he'd be picked up for cremation.

~~

I've no doubt that Rambo and I were meant to come together; he remains the greatest being I have ever known. Because he *insisted* on freedom, we gave it to him. I am a better person, and CAS a better sanctuary, for what he gave in return.

So here's to you, my boy. May you live forever in the hearts and spirits of all who pass through these sacred grounds. May you live forever in every blade of grass, every willow leaf, and in the gift and promise of each new day. And may your noble life, spent watching over all creatures great and small, human and animal, remind us all to make our lives matter.

Lumpy Sans Leader

David and I returned to the house, showered, and headed to New World Home Cooking for dinner. We sat at the bar; I order a gin and tonic, the first I'd had in years. David ordered a draft beer. Ditto.

"How are you doing?" he asked. He was the first of many over the next two days to ask the same question. I appreciated my friends' recognition of enormity of the loss.

"Absolutely fine," I said, because I *was* doing fine. Not only was I fortunate to have a long time to prepare for Rambo's passing, but with plenty of experience saying goodbye to beloved animal friends, I've learned that I grieve on the front end. Even with the Great Dog Murphy, director of Canine Pursuits at CAS for thirteenth and a half years, who shared my bed, laid outside the shower when I was in it, accompanied me everywhere I was allowed to take him and plenty of places I wasn't . . . Somehow, the sadness happens before the animal is gone. That was certainly the case with Rambo. Once the great beast was euthanized, surrounded by his two and four-legged CAS friends, I was at peace. Truly. Reverent, humbled, profoundly grateful for his friendship and the life-changing lessons offered during his eleven years with us. And after a difficult night anticipating the loss, I slept soundly . . . even though I was on a mattress in my yard sleeping with Lumpy, an ancient Merino sheep, who stood over me, *baaahhhing* balefully because he had lost his leader and was terrified. But we'll get to that in a minute.

In the last two years of his life, I shared Rambo's story with thousands of visitors. We had dozens of quiet moments alone at night in the barn, and up until six months ago, he was still able to perform his self-appointed job as guardian of all living beings at Catskill Animal Sanctuary. We'd had the gift of a long and rich and memorable goodbye.

Rambo's last act, in fact, was perhaps the most heroic, and therefore the most memorable of all. I was washing dishes early one morning before staff arrived. I looked out the window, and saw the ancient sheep limping toward my house.

"Mmmaaaaah," he called to me. "Something's wrong."

In that instant, a Holstein steer named Doc careened past Rambo, back legs kicking up everywhere in that goofy inimitable "I'M REALLY HAVING FUN!!" way. Great. The cows were out; I was alone.

I pulled on my shoes and rushed down to gather the herd— Bridget hiding behind a willow tree, Tucker trying to break into the kitchen, blind Helen in the middle of the parking lot, just standing there—but not before kneeling down in front of our guardian, looking into his knowing, golden eyes, and saying "Thank you, Rambo. Thank you for telling me about the cows."

My relationship with Rambo was special for many reasons, not the least of which was because I was the only person on the property from dusk until dawn, and Rambo knew this. We were a two-member night rescue team. The courage and nobility in his final act of leadership was in his willingness to ignore his own pain to let me know about the cows. *He was sixteen.* His shoulder hurt, his knees were knobby from arthritis, he rarely left the barn, and when he did, it was only to take a couple steps outside into the sunshine. Yet on that early autumn morning, he stood up from his plush straw bed, and hobbled five hundred steps or so to do his job as guardian of the CAS flock.

"Mom!" he called. "The cows are out!"

Thank you, Rambo.

～～つ

David and I pulled down the driveway at 8:30. "Let me check on Lumpy," I said. David dropped me at the barn and returned to the house. Inside the darkened barn, in the middle of the aisle where he'd slept with Rambo for the last several years, I watched as a terrified Lumpy reacted to every unfamiliar sound. His breathing was elevated, his crying incessant. As a member of the Underfoot Family, Lumpy hovered over Rambo for years as Rambo lay in his bed of straw. It always appeared that Lumpy was watching over Rambo; now it's clear that the opposite was true.

"Mm-aaaaahhh!" he cried, rushing to me. "*Maaahhh!*"

"He's gone, Lump," I whispered to the old sheep, kneeling beside him. I wondered how long it would be before it was Lumpy's turn. "I know. It's hard."

I walked from stall to stall, checking on the animals. "Hi, friend," I said to Noah, offering a piece of carrot. His neighbor Star stretched her neck out for a piece and I obliged. "Mmmaaaaaah!" Lumpy insisted, jerking around to see what monster was lurking behind him. Normally a reserved sheep, he was nearly touching my side, pleading for help. The unthinkable had happened. Lumpy had lost his leader. His world had come unglued. Sheep, you see, have incredibly strong flocking and following instincts; they run from what frightens them and band together in groups for protection. This "safety in numbers" behavior is the only defense they have from predators.

It's one thing to have this knowledge about sheep; it's something altogether different to see an animal, deprived of that security, come apart at the seams before one's own eyes.

I wondered if he'd feel more secure in a stall with goats Allen and Arthur. He had shared a stall with Arthur in the past. I opened the door, and moved to a back corner of the stall, holding Arthur and Allen by the collars to allow Lumpy room to enter should he want to.

"Wanna come in?" I asked. Normal sheep respiration is twelve to twenty breaths per minute. At that moment, Lumpy's was about one hundred. He stood at the opening for a full minute or more, rushed into the stall, up to me, and rushed out.

I sensed what I needed to do and knew that I was in for a long, long night.

"Good night, animals! I love you!" I called a little more softly than usual to horses Casey, Bobo, Beyond, Abby, and others; to pigs Jangles, Charlie, and Chopper; to turkeys Dierdre and Declan; goats Arthur and Allen; and to all the others in the Big Barn who missed their nightly check-in because their friend was far more needy than they.

And then, to honor him, I called out the words I've called thousands of times before leaving the barn after night checks:

"Guard the barn, Rambo!"

"You comin', Lump?" I asked him. Sure enough, the normally slightly stand-offish sheep was no more than a foot from me as we made our way up the driveway, past Julie's office, past the beautiful garden already sprouting beans, mustard greens, lettuce, and strawberries, and into my back yard.

"David!" I called softly, hoping he could hear me through open windows. "David!"

David walked out onto the back deck. A curious Hannah, my young black Lab, peered through the glass door. We sat for a few minutes. Lumpy wandered a few uncertain feet, looked around quickly, looked back at me. "Maaaahhhh!" We locked the gate to my small back yard, sensing that Lumpy would feel more secure if he were enclosed rather than free.

We sat quietly for what felt like a *really* long time. Lumpy turned in circles, the way Rambo used to do when he sensed danger. I suppose it's what all of us do when we sense imminent threat from any direction. Gradually, he settled. *Maybe he'll have a peaceful night out here knowing I'm right nearby*, I thought to myself. It was late, and I was wrung out from a brutal week.

As I stood and walked up the six stairs to my deck, Lumpy scaled the stairs and, quicker than I could blink, was in my living room. Hannah the dog jumped off the couch, not quite believing her eyes, and Lumpy fled, bleating even more frantically than before.

It was obvious that there were only two choices here: 1) take Lumpy to the barn, lock him in, and make him spend a terrified night alone; or 2) sleep with Lumpy.

"David," I sigh, because good *lord*, I was weary. "Will you bring the twin mattress out into the yard?"

He did, then he and Hannah headed back to his house in High Falls. And on the night that we had both said goodbye to our Rambo, Lumpy and I spent the night together, under the stars.

$\mathcal{D}uh$

In the summer of 2012, an international group of brain researchers released a document called the Cambridge Declaration on Consciousness. The document declared the group's data-driven consensus that most animals are conscious and aware in the same way that humans are, and confirmed that virtually all animals have at least some degree of sentience—even bees, according to Christof Koch in his *Huffington Post* blog, "Consciousness is Everywhere."

It boggles my mind that stuff like this makes headlines. All around the world, those of us who live and work with animals, *regardless of species,* utter a loud, collective *"DUH"* whenever science gives the tiniest of nods to what everyone else has always known. What's headline-worthy is not what science has "discovered," but instead: 1) why science is so far behind the rest of us and 2) how human animals treat non-human animals, given that plenty of us know that, human or non-human, we are all essentially the same.

What would the world's reaction be if the *New York Times*' lead story tomorrow were "Chickens Understand That Their Throats

Are About to Be Slit" or "Horrific Confinement and Deprivation Feels Same to Pigs as It Does to Humans"?

Take a look at what we witness at Catskill Animal Sanctuary:

1) It was time to euthanize an old steer named Samson. We had rescued him from a horrid hoarding situation. He and I had been special friends, but the time came when the old steer was no longer able to stand. Many of the humans who had loved him surrounded him, rubbing his massive body, singing. I sat at his head, and as he was falling asleep from the tranquilizer (the first step in a two-part process), Samson licked my face over and over—thirty times, perhaps?—until he could no longer hold his head up. My unscientific view? He was saying goodbye, saying thank you, and saying he loved me.

2) I've already discussed the wonderful friendship between Barbie the hen and Rambo the sheep. Barbie loved to rest on top of Rambo's back or cuddle up next to him in a pile of hay. When Barbie returned to the barn after a two-week illness-related isolation period, she walked out into the long barn aisle, glanced around, walked past several chickens, past the humans, right up to Rambo, who was resting in the aisle, and pressed her big bird body up against his. He turned his head, and gently nuzzled her back. There were dozens of moments like this between Rambo and Barbie. For me, that single one was "proof" of their affection.

3) A former cockfighting rooster named Paulie nearly always chose to eat lunch with staff. Alex always brought a small bag of sunflower seeds, and after he shared some once or twice with Paulie, the bird began walking immediately to Alex. If Alex didn't immediately produce the seeds, Paulie

pecked Alex's leg, looked up at Alex, pecked again. If Alex *still* didn't deliver the goods, Paulie became irate, squawking and flapping his wings, unwilling to accept no for an answer.

4) My back deck affords a wonderful view of the sunrise. I go out frequently to enjoy the experience—and so does Franklin the pig. Just as the sun is about to come over the cliff, Franklin walks out of his barn, faces the spot where the sun will momentarily rise, and waits. His peaceful anticipation feels identical to mine.

5) Lambert the sheep is new among our free-range crew. The other day, he was trying to befriend Lucy the cat. He approached slowly, lowered his head to hers. She whacked at him, hissed, and moved a few steps away. Lambert patiently persisted. "I want to be your friend," is what I felt him offering her—but Lucy wasn't interested and eventually gouged him on the nose. Lambert walked immediately to me and buried his face in my chest.

6) Every June, CAS hosts and event called the Shindig. Hundreds of folks attend; consequently, many witnessed Ethel the turkey walk purposefully out to the stage and sing for most of the day with the band. What was most striking was that, when they weren't playing music, she didn't sing. But every time the lead vocalist began a new song, Ethel was about two feet in front of him, singing along.

7) I was in one of our pig fields and honestly can't remember what I witnessed, but will never forget that Franklin the pig and I appeared to be laughing simultaneously. A "PAY ATTENTION TO THIS MOMENT RIGHT NOW" space opened up in my gut the way it always does when I'm witnessing something that challenges an assumption I hold. I

high-tailed it to the break room where the staff was having lunch. "Do you guys think pigs laugh?" I asked. They might as well have said, "Do you live in a HOLE?" In fact, one of them said something like, "That's like asking if we think pigs eat. OF COURSE THEY LAUGH!"

We humans know that dogs and cats are sentient not because science has "proven" so, but because we live with them and understand the subtleties of their behavior in the same way a parent knows those of her child. The same holds true, of course, for humans and farm animals, wildlife, reptiles, marine animals and exotics: those few of us whose job it is to encourage them to thrive know who they are, with their individual quirks, their rich emotional lives, their yearning to experience joy. When we read that brain researchers have finally acknowledged what is blatantly obvious, questions about politics, agendas, problems with methodology, who funds the research, and the like come to mind. More to the point, however, is the question of why we need science at all for the purpose of "proving" animal awareness. People around the world tell stories like those above. We need a means for anecdotal evidence to drive policy and practice.

My fervent hope is that one day soon, using the term "animal" to justify *virtually any horror humans want to inflict non-humans* will be as unacceptable as using race, religion, or gender. In the meantime, I take some small solace in the fact that science may be catching up with the rest of us. The fact that it now acknowledges the sentience of nearly every animal alive is one more tool in our belt when we entreat our visitors not to eat our friends.

PS. In the time that it took me to write this single essay, the USDA reports that almost one million chickens, 28,526 turkeys,

23,027 pigs and many thousands more animals—animals whom brain scientists recently acknowledged are conscious and aware, just like humans—were killed to feed us.

When We Smile

When the going gets tough, the tough kiss pigs . . . or horses . . . or sheep. Seriously. We do. We return to the joy of the work in this section, folks. Even though in one essay we are saying goodbye to yet another iconic animal, there is beauty and pure, raw power in our farewell. He was one of the funniest animals I've ever known. So the memories of his years with us? Pure, unadulterated joy.

Enjoy these quick snapshots; brief moments in time in the life of Catskill Animal Sanctuary. This part of the book includes one final, wonderful memory of the Great Sheep Rambo titled "The Audacity of Love." I chose to close this edition of the book with it because it brings home the point I've tried to convey throughout the text: We really are all the same.

I hope this book has convinced you of that.

Nine Happy Horses

My house is situated high on CAS property in the middle of a horse pasture. From my back deck, I can see the four cow pastures; one sheep pasture; four of our eight horse pastures; the duck pond; the "special needs" area for our blind duck, Sassafras, and his protector, Succotash; one of our rabbit houses; the turkey barn; the pig paddock; and three of our chicken houses. From my office window, I can call out to our blind cow, Helen, and her devoted seeing-eye calf, Dozer, who, at 6:10 on a frigid April morning, are still snuggled in their barn.

Indeed, I'm a lucky woman.

The sky lightens slowly. Beyond the pond outside my back door, sixteen cows rest peacefully at the far side of their pasture. Only young Jesse stands near the pond. At three years old, he's still filled with wonder, and he's fascinated by the wild turkeys that strut and preen and peck the ground in front of him.

Athena, Fritz, and Abby stand at my deck, staring at the door. Six other horses—Mango, Mary Jane, Hazelnut, Callie, Eloise, Katydid—turn from the pond's edge to join their pals. Still in their fuzzy winter coats, the horses nonetheless look good, and I'm pleased. They've all gained at least two hundred pounds since their rescue from a Saratoga horse breeder who admittedly "just didn't want to feed them." *How hard is that to understand?* he had asked me when we spoke after concerned neighbors were unable to get police to intervene.

I smile to myself, stunned that after just three days in this particular pasture the horses already have my number. They know that all they have to do is ask (and they're doing it beautifully by simply crowding the deck and *staring* at my door) and I'll emerge, treats in hand.

I take a five-pound bag of carrots from the fridge. "Good morning, girls and boy," I say. (Fritz was the lone boy from this particular animal rescue.) I sit on the deck and a pile of rescued horses surrounds me. I'm struck by their patience and politeness as they wait for me to open the bag and dole out the orange prizes. There's no jostling or competition—even Athena, the head honcho, allows others to cluster more closely than she. Beautiful Abby—pure white—nuzzles the top of my head as she waits. Abby is a wonderful success for Catskill Animal Sanctuary. Near death when she arrived, she could also barely walk: her hooves were a foot long and riddled with abscesses. It took more than an hour for her to limp, one painful inch at a time, off the trailer when she arrived. But here she is just three months after her rescue, galloping from one end of her pasture to the next, no sign of pain.

The horses munch their treats as I tell them how fine they look. They're grateful to be here. While skeptics or cynics or obtuse scientists would say I'm being anthropomorphic, they'd be *so* dead wrong. Rescued animals show their gratitude in myriad ways obvious to anyone who knows and observes them.

I wait for them to saunter off. They'll head toward the barn, knowing that it's nearly time for breakfast. Sure enough, after a few minutes of pats and praises, Athena turns. She has heard April exiting the kitchen with the breakfast bowls. Eight other horses eagerly follow, and another day begins at Catskill Animal Sanctuary.

The Underfoot Family

I n my lifetime, I've probably been in well over a hundred barns. My dad's farm had five barns. I took riding lessons at many different farms throughout my childhood, and I often accompanied my dad to various farms and racetracks. I've been inside the barns of many other rescue organizations, those of people wishing to adopt animals from us, and those from whom cruelty agents are seizing animals who are not being cared for. I've been inside plenty of barns that exuded misery and suffering and, thankfully, many that were filled with joy. But I've never been inside a barn like the main barn at Catskill Animal Sanctuary.

Take what's happening right now. I'm sitting in the middle of the barn aisle, trying to write about the Underfoot Family. Only I can't, or at least it's difficult, because Barbie the hen is insisting on getting in my lap, and her boyfriend Rambo the sheep—resident wise man, security guard, and the *real* boss of the joint—keeps pawing my thigh with his hoof because he wants a butt massage. If today is like most days, I'm sure he's already gotten, oh, at least a dozen butt rubs, but here I am, sitting in the aisle, another mark.

(Obviously, I'm the boss only in theory, just as Kathy Keefe is our theoretical farm manager and Julie Barone our theoretical business manager. One doesn't need an organizational chart to understand the hierarchy here; it's clear who runs the show.)

I am being swarmed by animals. Arthur the goat is nibbling my hair, and Hannah, Lumpy, and Aries, our other three free-range sheep, are lined up in front of me like three panhandlers. Charlie the pig walks past. He's on a food-finding mission, the mission shared by pigs worldwide.

Our newest free-ranger is a horse named Casey, the going-on-forty-years-old pinto gelding who arrived years ago after our concerned farrier told us of a horse living in a New Paltz junkyard. Casey, who survived the death of his best pal, Junior; Casey, who survived an abscess that swelled up like half a honeydew when he tore his cheek with too-long teeth; Casey, gentle giant, friend to all who call CAS home. Our low-lying pastures are the only ones suitable for Casey's old, arthritic joints, but in the spring, the mud is simply too hard on him. So now, each morning, Casey's door in the middle of the barn is opened, and he heads out—sometimes going left, sometimes right, sometimes immediately to the hill in front of the barn to graze, other times to visit the horses in the large pasture behind the barn. Casey visits the cows; he naps under the willow tree next to a pig or two. After a particularly full day, he often settles in the middle of the barn aisle in the midst of the smaller free-range animals until his eyes grow heavy with sleep. Yesterday, he was relaxing in my yard, and if I had encouraged it, I've no doubt that he would have walked right into my house. Casey delights in this life.

Why do we allow what many folks would frankly consider absurd? It's a legitimate question. Having a host of free-rangers doesn't exactly lend itself to *efficiency*. You try wheeling a cart stacked high with food dishes down the aisle and see how far you get, or try moving the tractor and manure spreader forward during stall cleaning when a sheep is standing in front of you and a duck wanders between the front wheels and two pigs are arguing on the floor to your left.

Before the driver can inch forward, she must wait for her assistant on the ground to shoo the gang out of the way and issue the "All clear!" signal.

Among the eight of us—Keefe, Alex, Jenn, Robyn, Russ, Erin, and me—there's not a single person good at setting limits. Alex loves to shout "Free the pigs!" as if a pig living in a huge heated barn with daytime access to a pasture that's even bigger was somehow imprisoned. If Russ had his way, every animal on the farm would be free to wander when and where he chose, and at least once a week, with Erin and Robyn's enthusiastic endorsement. Jenn and Keefe have added another being to our growing roster of free-range animals. The latest is a goat named James Curtis, or Jimmy for short. When Jenn, our animal care director, noticed that he was consistently losing weight, she was concerned about cancer or another grave illness. But all tests came back negative. Far more than we realized, gentle Jimmy was being ostracized at feeding time. He is a new member of the family who joins Arthur, a goat the size of a moose, as our second official greeter.

Its inconvenience to humans notwithstanding, I've always insisted on "the underfoot model"; CAS seems to draw people who revel in it. The belief that every living thing deserves happiness informs all that we do at Catskill Animal Sanctuary. In fact, our reasoning and our policies attempt to go even farther than that. Simply put, we do all that we're able to ensure that every single animal in our charge can truly *thrive*. And for a host of reasons, the quality of care we provide to most—large spacious pastures, generous barns and shelters, a superb diet, friends of one's species, and plenty of attention from humans—isn't enough for all. Blind animals, old animals with a variety of old-age afflictions (most typically arthritis), timid animals who need more frequent attention, "broiler" chickens (the term given by the poultry industry to chickens raised for meat) with industry-induced mobility issues, animals who are being picked on by their herd or flock because they're old or small or timid or all three—these are the ones who join the motley crew known affectionately as the Underfoot Family.

If there remains in any of us vestiges of ignorance or species-specific prejudice (for example, sheep are shy, turkeys are stupid), these animals are in our faces each day to challenge those prejudices. And regarding anthropomorphism, if you buy that concept, if you believe that this entire book is anthropomorphic, I challenge you to come to Catskill Animal Sanctuary, to sit quietly for half an hour among various members of the Underfoot Family, and to tell me at the end that we're attributing "human" emotions to animals.

Right now, for instance, it's extremely difficult to type. Why? As I sit in the middle of the barn aisle, laptop on my lap, Ethel, one of our free-range turkeys, is lying next to me, her feathery body pressed into my side. As I attempt to type, so does she. She's enthusiastically pecking the keyboard.

"Oh, I see you've got a few assistants!" Keefe offers as she walks out from the kitchen, her arms stacked with feed dishes for some of the smaller animals.

"I don't need your help, Ethel," I say to her; evidently her hearing is selective, because she continues typing. Across the aisle from us, Zoey and Charlie, two potbellied pigs, are arguing ferociously over who is going to get the favored spot in their enormous bed of hay: both always want to be closest to the wall. Meanwhile, Hannah of the Three Wise Sheep moves forward until she is two inches from a turkey named Agent Forty-Four, at which point her pals Aries and Lumpy turn and head outside to discover what awaits them.

I was going to make a point about the closeness of this family, about the seamlessness of their days, about what a privilege it is to observe their adventures, their kindnesses, their unequivocal acceptance of each other, their schemes (nearly all of which have to do with breaking into the kitchen). I wanted to discuss how much love they give to us, and to each other, many times a day. And I wanted to speculate about how many vegans they've created, for to be among them is to understand that in the ways that truly matter, we really are all the same.

However, the Underfoot Family is making anything that I'd like to do extremely difficult! Casey the horse has just wandered in from his visit with the cows and has joined Hannah the sheep, Ethel the singing turkey, and Petunia the pig, who is pressing into my knee, then my thigh, with her snout. So here I am, surrounded by animal friends who could be anywhere they choose but have all chosen to join me in the middle of the barn and assist in the writing of this essay. And though I do apologize for needing to end it so abruptly, I trust that you get the picture.

When Winter Kicks Your Ass

An Important Note:

Keefe, Jenn, Robyn, Alex, Russ, and Erin, as well as intrepid volunteers Donna, Dawn, Jill, Anthony, Phyllis, Julie, Jackie, Sue, Christine, Darcy, Lisa, and Carol would kick my skinny white behind if I didn't explain here that this essay is an anomaly. *They,* not I, are the ones who freeze their tushes off on winter days. I fill in on occasion, and work on the farm a few hours a week whenever I am able (my spirit needs it!), but any one of them could have written this chapter from firsthand experience. No matter the conditions, their job is to show up on time and get the day done with a smile. Hats off, and thanks, to all of them, and enjoy this essay that features the "back-up crew" of Caleb, Michelle, and a former employee named Zach.

It's 6:30 a.m. when I enter the barn to set up feed. An additional six inches of snow has fallen overnight; the white stuff is well above my knees. A dangerous layer of ice glazes the ground below it. Willow branches coated with ice glisten in the sun. It is another breathtaking, beautiful, treacherous day in the worst winter any of us can remember.

Around the globe, scientists, doomsdayers, and conspiracy theorists are speculating about the causes of mass deaths of cows, songbirds, buffalo, fish, and other animals. Theories range from the plausible—extreme cold—to the prophetic—the end of the world is upon us. I, for one, would put my money on the toxic waste emitted into our waterways, soil, and air by agribusiness and hydrofracking and would also bet that if these industries (or other corporate giants) are the culprits, the public will never know. Right at this moment, however, I have more mundane concerns.

I pull feed bowls down from the shelves by the dozen and group them on the concrete floor. I turn on the water (no frozen pipes!), then move to the medical shelves to grab various supplements: glucosamine, "bute" (an anti-inflammatory), probiotics, and so on. Seven large bins labeled chicken feed, goat and sheep feed, senior feed, etc. line the wall; many are subdivided inside. We believe in giving our animals the best care possible, and that belief demands customized diets and a feed room as complex as a professional kitchen.

By 7 a.m., staffers Michelle, Caleb, and Zach have arrived. "Thank you *so* much for getting here early!" I say effusively, knowing that folks driving from greater distances will be late. Caleb heads out to plow; Michelle backs two trucks to the hay room to load them with hay for our two outside feed routes; Zach stays behind to help me feed the sixty or so animals who live in or next to the main barn. So far, so good.

But by 7:10, the plow truck is fender-deep in a snow bank, and by 7:15, the tractor that was supposed to pull it out is stuck as well. We take the Chevy pickup, our workhorse, down equine alley, hoping to retrieve both the plow truck and the tractor. Unfortunately, it is already loaded with hay and grain, so the animals on our "barnyard" feed route will have to wait for breakfast.

At 7:20, we back the watering truck up to the barn, pull a long hose toward it, clip the hose to the two-hundred-gallon cylinder that holds water for our outer fields, and wait. Nothing happens. The hose is frozen. So at 7:25, we unhook the hose, step down from the truck, walk the hose halfway down the long barn aisle and into the heated kitchen where it can thaw, along with Wallace the rooster, who was shivering when I looked in on his flock a few minutes earlier but is now perched in front of the heater with outstretched wings.

We do our best in these conditions. Still, by 8:30, the snowfall is getting heavier by the minute, two of us have slipped on ice, three vehicles have gotten stuck in snow, and a hay delivery has been postponed for the second time due to weather. We'll have to pay through the nose to get one hundred bales, a two-day supply, from a local retailer.

For safety reasons, we decide to keep our special needs horses—blind, ancient, or both—inside for the day.

"I wish all I had to do is curl up in a warm stall and watch winter happen," offers a volunteer named Christine, crazy to show up on a day like this.

"Yeah," says Michelle. "I should be at home on the couch with my four cats." She has a smile on her face nonetheless. Somehow, we all do.

Sanctuaries devoted to the well-being of the animals in their charge are 365-days-a-year operations—no matter what the weather. There's no such thing as "I can't make it in to work." Days like this one begin early and end late, the length of the daily "to do" list doubles as sub-zero temperatures and a restless sky conspire against us. Among the daily "extras" are blankets for the horses; extra bedding for pigs, chickens, goats and sheep; the shuffling of animals from outdoor shelters into our large main barn; and the constant monitoring of our vulnerable animals. Of course, animals are far better equipped to withstand cold weather than we are, but −15 degrees makes me nervous. Ice must be broken, locks must be thawed, doors are frozen shut. Our fingers and toes ache, as does the skin on our faces. It's so cold that it hurts to breathe.

"We're getting up to thirty inches on Wednesday," announces Caleb, our resident jokester. Somehow, we all know that this is not a joke.

"Are the animals okay?" people want to know. In fact, in this weather, we get this many times a day. Sometimes, it's when I'm in line at our local health food store. Most often, volunteers and supporters email us; occasionally, we receive phone messages from folks we've never heard of.

So let me make it official: Even when winter kicks our human behinds and knocks the breath out of us, the animals are fine. We make sure of that.

These are exhilarating, exhausting days. These are days that test everything from our tractor batteries to our mental fortitude. We are truly bone-weary. But, somehow, we manage to laugh our way through this epic winter, fortified by our respect for each other, our devotion to the critters who call CAS home, and by the animals, who lift us up and cheer us on.

Really? you ask. Yes, really! When Franklin the pig looks up from a fluffy bed of hay and grunts a "thank you," or Abby the horse gallops through the snow to greet her friends whom she's missed overnight, or Henrietta the turkey careens out of her cozy stall each morning because *she just can't wait to say hello*, they lift us up, cheer us on, prop us up for one more round in the worst winter any of us can remember.

Pig Kissing 101

"So sorry about the delay," I say as I approach the twenty or so guests waiting patiently under the willow tree for the next tour to begin on this crisp October afternoon. "We're missing a tour guide today." I tilt my head back for a good long drink of the water and juice mix I carry with me on tour days. It's going to be another day without lunch.

I take a seat on one of the benches, but just as I'm about to begin my introduction, Franklin the pig looks up from his spot in the far corner of the pasture that borders our waiting area, presenting an opportunity that can't be ignored.

"Kids!" I summon the group. "I need your help!"

I point to my friend with the pink skin and fuzzy ears, then say, "See that pig over there? His name is Franklin, and he loves children. Right now he's having fun digging in the dirt, but if you help me, I bet he'll come over."

"Do we have to go get him?" asks an outgoing little girl of six or seven.

"Nope. We just have to call him over. I'm going to count to three, then I want everyone to shout as loudly as you can, 'Franklin!'"

As his name leaves our collective lips, Franklin leaps the creek that divides his field in half and is trotting in our direction, grunting in anticipation. The group hovers around me.

"Whath he doing?" a wide-eyed child of five or six lisps through the hole in his top teeth, uncertain whether to laugh or to flee in terror from the seven-hundred-pound body barreling right at him.

I squat so that we are eye to eye. "What's your name, sweetie?" I ask of the little boy, who by now is nothing but breath and bulging brown eyes.

"Malcolm," he whispers, glancing furtively at his mom.

Franklin is now a mere foot from us, pushing a soft snout into the wire mesh fence, his requests for company growing louder by the second.

"He's talking to us," I explain. "He's saying, 'Malcolm, come right here so I can meet you. I bet you'd be a great friend.'"

"Thath really what he'th thaying?" Malcolm asks.

"Abso*lutely*!"

Malcolm smiles.

"Hi, everybody," I say to the group. "I'm Kathy Stevens, founder of Catskill Animal Sanctuary. In just a moment you'll learn about the mission of CAS—who we rescue, how we make our choices, why we encourage all our guests not to eat animals like my friend here. But right now, we've got some pig kissing to do."

(Sadly, as a result of the swine flu scare, this frequent ritual has been removed from our weekend tours. Many of the staff, though, still kiss the pigs, along with all the other critters. I rarely go a day without kissing my special friend Franklin.)

A few chuckles emanate from the group, and one woman says, "I've been looking forward to kissing a pig all summer."

I edge over until I'm right in front of Franklin and offer my hand to Malcolm. He takes it.

"Pigs are loud, Malcolm, and that's a scary thing if you're not used to it. But look: Franklin can't come any closer because he's behind this fence," I explain, touching the top rail of Franklin's pasture.

"Sit right here," I encourage him, and little Malcolm folds his legs and sits so that our knees are touching. "Hi, best pig in the world. Hi, you good, good pig," I say to my friend as I flatten my hand against the metal mesh so that he can push into it with his muddy snout the way he likes to do. "I love you, Franklin."

I take Malcolm's hand and hold it beneath mine, and watch the child's smile grow as Franklin greets him.

"He'th all muddy," Malcolm giggles.

"Yeah," I say. "Pigs need mud! It helps them stay cool since they don't sweat, and it helps prevent sunburn and even protects them from pesky flies. Besides, you might be just as muddy as Franklin when you leave here!" I glance upward at mom.

"Not a problem," she responds, her smile as big as her son's. "This is worth a little mud." The group has gathered around us, and I sense another opportunity.

"Well, everyone," I turn around to address the group, focusing on the children. "I haven't given Franklin a kiss yet today, so I'll be right back."

I hoist myself up and over the fence and step down beside my porcine pal. Franklin rubs his cheek against my thigh and oinks his most emotional hello. I talk briefly about the wide variety of pig vocalizations and how easy they are to understand.

When I kneel, I am smothered in pig kisses: wet muddy snout against nose, cheek, head. I kiss him back, then smile to the group. Most are laughing with delight, except for one woman who looks like she wants to grab her child and flee from what is surely a demonic cult. ("Carlton, *they actually kiss pigs*," I imagine her saying to her husband over their pork chop dinner. In fairness to her, I recognize that what she's seeing is a little, uh, unorthodox.)

"Anybody else want to kiss a pig?"

Before she either faints or vomits, the pork chop eater does, indeed, take her child and head toward the parking lot. In the meantime, two young girls are squealing with glee, entreating their parents.

One at a time, Dad passes each of the girls over the fence. Franklin, of course, is beside himself, and the girls are instantly both filthy and in love. "I love you, Franklin," the older one says. "I love you, Franklin," the younger one mimics through delighted giggles as a cool snout greets her.

I pass the human packages back over the fence. Malcolm, frozen in place on the other side, looks up at me, his eyes saying everything.

"Okay, trooper," I smile as I hold out my hands to help him over. "Ready to have some fun?" He looks at his mother, who asks simply, "What do you think, Malcolm?" Malcolm utters something inaudible as he stands and raises his hands to be lifted over the fence. Mom passes him to me.

At first, Franklin's mass, presence, and proximity overwhelm him. He stands frozen, his back to the forward pig and his face buried in my shirt. Franklin pokes the small of Malcolm's back, and a muffled giggle emerges. Franklin pokes some more: Malcolm's back, his fanny, his legs. The little boy's giggle becomes a laugh, his laugh a guffaw as Franklin ramps up his enthusiasm.

"Franklin, you're tickling me!" Malcolm shrieks. By now he has turned, and Franklin is laughing deliriously with him, that open-mouthed pig laugh that, once witnessed, is not soon forgotten.

"You didn't *tell me that pigth were ticklerth,*" Malcolm cried.

"I know I didn't, Malcolm! Franklin wanted to surprise you!" I responded.

I glance up at Malcolm's mom. A smile as wide as the Mississippi is spread across her face.

Home for the Holidays

Thoughts on Easter Sunday

In the midst of such grandeur, there is the wretchedness we
have wrought.

David and I have just returned home from vacationing
in Utah's Zion National Park. It was a glorious week in God's
country, with the exception of one moment that will be seared
into my memory forever. It haunts me now as I walk the grounds
of Catskill Animal Sanctuary on this Easter Sunday.

As we set out on the first of what would be epic daily hikes,
David and I spotted a small pen filled with horses and mules tacked
up for the day's trail ride. While I am *far* from being a fan of these
"rent a horse" outfits (the animals lead a life of drudgery, at best, and
when they've outlived their usefulness, most are sent to slaughter),
these animals actually looked terrific: Their weights were ideal,
their hooves were in great shape, their coats were healthy. I was
also heartened to learn from the guy running the outfit that the
mules and horses work for two days and then have a day off. Many

"dude ranches," trail rides, and carriage-horse operations give their animals only one day of rest per week.

"They're luckier than most," I said to David.

And so we headed out, exploring the glorious Zion Canyon, spending an absolutely delightful afternoon, until we got to the fork in the path.

"Let's go this way," David suggested.

To our surprise, after a few hundred steps, we arrived at the barn where the horses and mules lived, and my fantasy of spacious pasture and ample shelter where the hardworking animals would enjoy their day off was shattered. More than forty mules and horses were packed into a turnout pen of about a thousand square feet. On their day off, these animals, mostly mule geldings, had only a tiny, dusty, shade-less pen. No trees, no grass, no room to frolic. The single run-in shed was too small to accommodate even a third of the animals, which presumably means that when the scorching sun is high in July and August and the thermometer reads well above a hundred, many animals must simply endure it.

I checked their water. It was green. Thick algae lined the trough and an oily scum floated on top. *Shit*. People call Catskill Animal Sanctuary when they see animals enduring conditions such as these. *Can you do anything?* they ask, naively hoping that we'll either take the animals, file criminal charges, or guide them through the process of doing so.

"No," we say, more often than we would like. Like anti-cruelty statutes throughout the country, New York State laws are too lax and too vague. Though some counties are far better than most, in general, law enforcement resists prosecuting all but the most horrific cases. If an animal's weight is good and water (even if it's filthy and not potable) and shelter (even if it's a tree) are available, no law is being broken. All one can do is look for ways to engage the owner about his animal husbandry practices and monitor the situation to ensure it doesn't get worse. It often does. Then we *can* encourage law enforcement to intervene.

I sat cross-legged in the dirt outside the paddock. Four animals—a mule and three horses—approached, their ears perked forward in interest. Three were old mares, their huge bellies sagging, their bodies scarred from the kicks and bites of more dominant animals. These animals were far too old and worn out to be ridden. I know their fate. Like most horses who outlive their usefulness to humans, these will likely be slaughtered.

"Hi, girls," I said as they stood inches in front of me. "I'm sorry I don't have treats." Still, one nuzzled my face and another lowered her head and allowed me to massage her cheeks.

Twenty days a month, year after year, these animals work hard, carrying mostly those who know nothing about riding—nothing about sharing an experience with an animal—back and forth along the same dreary path, in temperatures that rise to 110 degrees. Their reward for their service? A few days off in a hot, tiny, dusty pen drinking contaminated water.

Across the United States, there are tens of thousands of such animals . . . as well as those in zoos, in traveling circuses and rodeos, at canned hunt facilities, at racetracks, and at theme parks. They endure their respective wretchedness because humans want to be entertained by them. So here's a question for you: Would the quality of your life diminish if you never again attended a rodeo, a petting zoo, a dog or horse race, a circus? Would your vacation be less enjoyable if, instead of riding a bored and overused animal, you and your family hiked the trails yourselves? If instead of taking the kids to SeaWorld to watch dolphins jump through hoops, you took them to the *real sea* to swim and build sandcastles? Before you participate in events that exploit animals, I invite you to consider them from the animals' point of view.

That human beings feel entitled to use animals for any purpose we determine to be fun or profitable is a level of callousness or obtuseness or disregard common to most of us in the developed world. For most, it's an unquestioned assumption of privilege; for a few, it may be a resigned sense that "it's just the way the world is." It never occurs to most of us to question the status quo. I'm asking you to do so. And I'm asking you to acknowledge your own role in the suffering of animals.

It's this very concept that Matthew Sculley discusses in his book *Dominion*. It's not a new book, but in its unflinching presentation of what animals endure at our hands and its call for mercy, it's one of the best for my money. Sculley asks whether "man's dominion over animals," as discussed in the Bible, suggests dominance or caretaking. Providing snapshots of the brutality and misery inherent in many animal industries, Sculley makes the obvious case that humans do, indeed, dominate all other animal species with breathtaking disregard of their suffering. Very few of us consider these issues. When we do, in my view, there's no argument to be made. We have enslaved the animals, plain and simple.

But this domination/subjugation/oppression model was not God's intention, Sculley argues. No, by giving us "dominion over animals," God intended us to be their caretakers.

Today is Easter Sunday. As I walk around this precious animal sanctuary on this symbolic day, joyful animals, allowed to be themselves, surround me. Policeman, a thousand-pound pig rescued from a Bronx apartment, is one happy camper stretched out on the sunny side of the shavings pile. Molly the cow and Sammy the steer playfully butt heads. I smile as Helen the blind calf licks the face of Andy, the young gelding still too weak to be turned out with other horses. No matter; Helen takes such good care of him. The goats play "King of the Mountain" at their rock pile, and, in one of the surest signs of the coming of spring, the hens are busily scratching in the dirt for whatever edible tidbit they can uncover. At the far end of the farm, Buddy the blind horse rolls blissfully on the cool ground. I watch these lucky few, and I remember the Zion horses.

Jesus suffered mightily. So do the animals. The power to change this is in our hands.

Happy Easter.

We Ate Pumpkin Pie

I spoke on Thanksgiving afternoon at the Berkshire Vegetarian Society's "Living Thanksgiving," held at the United Methodist Church in Lenox, Massachusetts. Roughly one hundred people were there; no turkeys died to feed us. Rather, there was table after table of everything *but* dead animals: cranberry sauces and pilafs, potatoes and yams, and entrées like loafs and soufflés and tortes and stuffed vegetables. And then, of course, there was stuffing . . . lots and lots of stuffing. Isn't Thanksgiving dinner really about the stuffing, after all?

My beloved pal Murphy the dog was invited in, and he had a grand time in the midst of so many good smells, so many animal lovers. Murphy begs very politely, and more than one guest succumbed to his patient entreaties.

I read from *Where the Blind Horse Sings*. I talked about the life-altering lessons learned from broken animals made whole again.

Questions from the audience were wonderful and provocative, and in the end I invited the group to visit Catskill Animal Sanctuary the following day. It was a lovely afternoon.

I pulled up to my house at 6:30 under a nearly full moon. I walked up the back steps, stripped off two jackets, and walked immediately to the oven with the pumpkin pie given to me as I left the church. I piled old newspaper, then cedar kindling, then locust and oak logs in the fireplace, and an amber glow lit the living room.

Just outside my front door, I heard the horses. I peered out to see old man Maxx and his friends Callie and Hazelnut. Their heads leaned over the deck railing; their ears pricked forward in eager anticipation of a friendly greeting and a treat. No carrots to be found, however, and my fruit bowl, often piled high, was empty. I wondered. . . .

I pulled the pie out of the oven, shoved one arm then two into my red corduroy jacket, then my green one, and walked out to the deck.

"Animals," I whispered, "animals . . . *look what I have!*"

Murphy got the first bite, shoving his snout right into the center of the pie. Callie was next. She sniffed tentatively. *What kind of treat is this?* she must have wondered! She tentatively licked the surface, ever the lady. Hazelnut did the same.

And then came Maxx. Maxx, the thirty-seven-year-old gelding surrendered to CAS after his owner died of cancer. Maxx, the gelding with his harem of six mares. Maxx, the pumpkin pie lover who took one perfunctory sniff and then smashed his muzzle so forcefully into the pie that I nearly dropped it, and then again, delighting in its texture and sweetness.

"*Happy Thanksgiving, Maxx!*" I exclaimed, laughing heartily.

Murphy and I stayed on the deck for another few minutes, surrounded by horses content to remain right there with us, drenched in moonlight, soaking up the love, and savoring pumpkin pie.

Merry Christmas, World!

I could be with David in Hawaii, my dad in Florida, my brother in Virginia, my sister and her delightful brood in Michigan, or my grandmother, aunts, uncles, cousins—the whole maternal clan—in Nashville. Instead, clad in long johns, jeans, boots, gloves, hat, T-shirt, turtleneck, fleece vest, and jacket, I'm scooping poop at Catskill Animal Sanctuary, and I couldn't be happier.

With our two animal caretakers either on vacation or taking the day off, I'm in the barn—per usual—on Christmas Day. WAMC, the public radio station, is airing Christmas essays, including David Sedaris's hilarious account of his single day working as an elf in a shopping mall. April and Allen and Alex are here with me. Quickly and effortlessly, we divide up the morning feed routine: April and Allen feed the "outside" animals, mostly big animals in big pastures the farthest from the barn; Alex feeds the "barnyard" animals—the rabbits, ducks, geese, chickens, and goats in seven different shelters clustered closer to the main barn; and I feed the menagerie inside the barn—eight special-needs horses whose age or condition have earned them a permanent spot there; the eighteen potbellies and big pigs who appreciate the heated stalls; twelve goats; Lama and Jack, our two blind (or nearly) sheep; an eclectic assortment of birds—five broiler roosters, Norma Jean the turkey, roosters Sumo, Rocky, Doodles, and Scribble . . . and so on. Today, a few extra treats are placed in each feed dish. Today, each animal gets a kiss.

The chickens get held, the pigs get massaged, every horse muzzle has a kiss planted on its smooth, warm center.

"Ummphhh," Franklin the pig grunts in gratitude. And Norma Jean, our rescued turkey, settles into my lap—uncertainly at first, but with each new breath, she lets go a little until her eyes are heavy and she's asleep.

I steal away mid-morning and an hour later return with three dozen pancakes. Christmas brunch in the barn! We pass juice and maple syrup, and vegan dietician George Eisman and his girlfriend, Melanie Carpenter, come by with one of Melanie's extraordinary desserts. So what if we've just finished a pound of strawberry pancakes apiece? It's Christmas! We dive into Melanie's chocolate mousse pie. This food is all made without animal products. And it's all divine.

Outside the kitchen door, Franklin grunts. "Can I come in?" he pleads.

We're tempted, but as you know, Franklin is no longer the five-pound piglet who arrived at Catskill Animal Sanctuary three winters ago. He is seven hundred pounds, and a seven-hundred-pound pig loose in a kitchen wouldn't be pretty . . . not even on Christmas.

I grab a handful of pancakes and slip out the back door.

"Merry Christmas, best pig in the world," I whisper to my friend, who gleefully gobbles the pancakes. "Come on, boy, it's time to go back to work," I say to him, and he and I head down the drive to clean the goose house.

Merry Christmas, world.

If I'd Had a Camera

I wish I'd had a camera. But then I'd have ruined the moment.

I've never once walked in the barn during work hours and not seen a human. Not once in all these years. But today it happened. Alex was up in the large hill pasture reinforcing fence. (Two mules arrived Sunday—their family's farm is in foreclosure—and one of them, Blackjack, is nicknamed Houdini. "Got a weak section, he'll find it," his owner explained. Sure enough, he had.) Meanwhile, the rest of the crew were cleaning the large cow barn at the back of the farm. Only I didn't know this.

Murphy and I walked into the barn. "Where is everybody?" I asked the yellow mutt, who trotted toward the kitchen halfway down the aisle in search of his pal.

Five feet from the entrance, Rambo lay in the middle of the aisle, holding court per usual. Beside him stood Agent Forty-Four the turkey, gently pulling bits of hay from Rambo's wool. Potbellies Zoey, Charlie, and Ozzi were there, too; surprisingly, they weren't searching for food. They were simply there, enjoying the company of their friends.

I plopped down with them. Murphy did, too, right by my side. To my delight, the animals allowed us to enter their peaceful circle—and just *be* with them. No one charged over to beg for food; neither did anyone walk away because a dog and a human had entered his space. Hannah, Rambo's woolly pal, the sheep found in a Queens cemetery, strolled over to nuzzle Murphy the way she always does. Mufasa the goat was with her. Above all of us, Maxx, the sweet old gelding who recently moved into Dino's stall, hung his big head over the four-foot stall wall, and there we were together: two sheep, a turkey, a goat, three pigs, a horse, a dog, and a human.

For a few precious minutes we sat . . . that's all. Miraculous simplicity.

But then Claude—the giant pink pig with the bad leg that earned him free-range status lest he be picked on by the stronger, more dominant pigs in the pig pasture—strolled in from the far end of the barn.

"Hey, big man . . . hey, sweet pig," I called softly to him. A few heads turned in his direction.

"Mmmph," he responded. And then he walked not into his stall the way he always does, but past it, straight toward us. He walked right up to Maxx the horse, his scratchy pink back nearly level with Maxx's muzzle, and he lifted his snout to Maxx and there they were, wet pink pig nose pressing into soft black horse nose. They stood there, pig to horse, Claude looking up intently, somehow knowing that Maxx would not lunge at him the way the horses typically do at the big pigs. The way Maxx, who truly can't stand pigs, always does.

Laugh, shake your head, call me anthropomorphic if you're skeptical or obtuse or disconnected or afraid, but I experienced what I experienced, and what I experienced on a cold winter day was pigs and turkeys and goats and humans and horses and sheep and dogs enjoying each other's company. Happy just to be.

Carpe the Diem

Dear Maxx:

There is nothing more that we can do. Five years you've had with us, and five plus thirty-two, the age at which you came, is thirty-seven. You're an old boy, Maxx. Very few horses get this many years.

Still, that fact is little comfort as I face your final day on Earth.

How I love you, Maxx! I struggle hard to breathe deeply, to gather the strength that I will need to send you off with a smile on my face a few short hours from now. I hope I'll be able to do that; I hope we all will. It's what you deserve.

Like the rest in this menagerie who call CAS home, you came to us because you had run out of options. Our waiting list is long and getting longer, as are the lists at sanctuaries and shelters around the country. Opening a thousand more shelters wouldn't put a dent in the number of animals desperate for a safe haven, desperate for someone to say, "You matter to me." All sanctuaries have to pick and choose, and when your guardian angel asked, "Who on Earth wants an old quarter horse with bad feet and a bum leg?" how thrilled we were to be able to say, "We do." We were her last phone call, she said. She had run out of options.

Yes, old boy, you mattered to us! A posse of people tried to help you after cancer took your human. They didn't want her final day on Earth to be yours, too, and their efforts to find a haven for you were truly Herculean. How stunned they were when I said, "Yes. We'd be delighted to give Maxx a home."

And so you came to this place where the unwanted come—a place where animals roam acres of pasture filled with rich grass, where they have spacious barns in which to bed down for the night, where they see not contempt but kindness in people's eyes.

Here, animals learn what it feels like to matter to someone—to *many* someones. In this sense, you were far ahead of the game.

To our delight, you were everything they said you were: smart and comical, trusting and unflappable. And while they didn't mention it, we humans all admired how accepting you were of the discomforts and intrusions of aging: the creaky bones, the missing teeth and soupy meals, the arthritic shoulder, the painful and recurring laminitis. When you coliced last summer, you were tubed *five times*—five times in four days, the vet ran a half-inch rubber tube through your nose, down your throat, into your belly, pumping you full of mineral oil to dislodge the impaction that just wouldn't pass through. Five tubings, four nosebleeds, but man, Maxx, you were a trouper. How very clear it was that you knew you needed help, clearer still that you weren't ready to go. When you finally pooped (six times in a single afternoon), we were overjoyed! Ebullient! Yes, it was selfish; none of us were ready to say goodbye.

You were so much more than "horse," Maxx. You were teacher, friend, playmate. Yes, playmate! With Hannah and Ozzi and Tucker the cow and madman Clarence the miniature horse

201

and Nutmeg the Invisible Hen in our midst, we certainly have our share of characters!

But none like you. Whether you were stretching your neck out to grab the edge of the wheelbarrow with your teeth to drag it closer for inspection, lunging at poor unsuspecting pigs passing by, or play-fighting with Ted through the stall window, you entertained yourself constantly and, in doing so, entertained us.

There's five hundred-year-old Maxx galloping at full-tilt the entire length of his pasture to ensure that Noah and Ted, the other two geldings, have no chance with *his girls*. There's Maxx with his head in the garbage can and oops . . . there's Maxx in the kitchen. There's Maxx, his head centered on a human's back, flinging it up and sending her flying across the aisle. Again. Again. Again. There's Maxx removing his support boots. And the final image: there's the brown streak of you, Maxx, dashing past Julie's office. The body ages; the spirit needn't. That's what you modeled for us.

How quickly you became our "starter horse." Yes—that was your job, Maxx! The tentative volunteer could enter your stall without fear, and when she was ready, you could reassure her as, instructed by humans who'd done this thousands of times before, she pulled the noseband gently over your muzzle, slipped the nylon strap first over your right ear, then over your left, then clicked the buckle and *voila*! Your halter was on, and she could lead you to your pasture, and you would be so very gentle, knowing, perhaps, that she was afraid.

You were the only one atop whose fanny I would cross my arms, resting my cheek on the crest of your tail, saying, "I'm going to kiss your ass, Maxx." I did, over and over, and you'd stand motionless, and together we'd prove the point that character is far more important than species in determining "how to be" around an animal—what behavior is safe, what's off limits.

These lessons and dozens more.

"Heather's here," says April over the walkie-talkie, and from various stations around the farm, staff and volunteers gather in front of your stall to say a gut-wrenching goodbye and honor you whom we all love. April and Walt, Allen and Alex, Abbie and Troy, Keefe, Chris, Anthony, Donna, Melissa, and Amy.

Heather injects your fetlocks with a nerve blocker so that you can leave your stall. It's simply too difficult to get a body as big as yours through a three-foot door. If you can walk out comfortably, it will be so much easier after you're gone.

And you do walk—exactly seven steps and no more. I walk backward in front of you, and when you stop, I drop the lead rope. It is your right, not mine, to decide which step is your last. You stop, not surprisingly, in front of your friend Ted's stall.

"You done, old man?" I ask.

Our beloved vet Heather O'Leary gently explains what's about to happen for those who've never been present for euthanasia. Other than her gentle voice, there is only muffled crying. We are steeling ourselves. We do not want this. *I do not want this, Maxx.*

"Are you ready?" I ask, looking into your knowing eyes.

I suspect I will tell the story of what happens next for as long as I live. You turn your head to the left, stretching your neck as far as it will go, to look back at your human friends who stand on that side. Your gaze lingers a moment on each of us, then you bend your neck around to the right, looking behind you now at Donna, at Chris, at Abbie. Such profound clarity in that unexpected final act of saying goodbye.

Thank you, Maxx.

You crumble to the ground as the anesthesia takes effect. Gently I lift your big head, scooting under it until it is in my lap. I rub your cheek as you fall soundly asleep. Heather asks if we're ready; she injects the drug that will stop your heart, and you are gone.

Sometimes, being silly, I'll say to Murphy, "Come on, mutt, we've got to carpe the diem!" and we'll dart out the door to whatever adventure awaits us. You did that, too, Maxx. Up until the very, very end, painful shoulder, sensitive belly, and bum feet notwithstanding, you carped every single damned diem—no, every single damned moment—you were given. It was your gift, your abiding lesson, for those of us who remain.

The Audacity of Love

That Hannah the sheep is in love with Rambo the sheep is no secret. Indeed, it's obvious even to first-time volunteers as Hannah bolts from her stall each morning in search of her Romeo. If she finds him immediately, all is well. But if Rambo is out of sight—either intentionally hiding or simply munching hay in a newly-vacated stall—she is initially disturbed, then worried, and finally panic-stricken, uttering a heart-wrenching, baleful *baa-aah* as the time it takes to find her soul mate increases. Once she locates him, all is again right in her world. She settles into her sheepness, content to roam the barnyard, grazing, stealing alfalfa from the hay room, and plotting kitchen break-ins . . . that is, as long as Rambo is no more than a foot or two from her. It is a relationship that she needs desperately, and one that Rambo some-times seems to appreciate, other times only tolerate.

Enter the other woman.

Barbie is a "broiler." As I've discussed in previous essays, this is the term used by the poultry industry to describe chickens intended for meat. Broiler: They exist for the sole purpose of being broiled.

Or baked. Or barbequed. She's one of hundreds who've arrived at CAS throughout the years from one of New York's five boroughs, lucky escapees from live poultry markets, slaughterhouses, transport trucks, and the ritual sacrifices of Santeria. We've taken chickens from dumpsters, chickens tied to trees in Central Park, chickens stuffed in mailboxes, and chickens who were drowning in crates left in flooding streets. Our latest, Barbie, was found in Brooklyn, hiding under a blue Honda.

Like Hannah, Rambo, and many more of our smaller animals, Barbie free ranges during the day. While she is young, the exercise is good for a body that will quickly grow morbidly obese. There's also no outdoor home for Barbie, as our ratio of roosters to hens is about 300 quidzillion to one. (Few people, unfortunately, want a pet rooster.) So Barbie snuggles into her home in the main barn each night, then each morning is lifted out to explore the barnyard and cozy up to whomever she chooses.

Unfortunately, just like the rest of the girls who've come before her, Barbie has chosen Rambo.

For several weeks, Barbie has been napping right next to Rambo, sometimes so close to him that surely even through his wool Rambo feels the heat emanating from her big bird body. Sometimes she climbs on top of his back, the patient Rambo motionless, and falls sound asleep. Rambo takes her overtures in good stride.

For a while, Hannah tolerated the new friendship. After all, Barbie was merely a hen; Hannah could still rest side by side with her love or stalk him relentlessly as he traveled the barnyard ensuring all was in order. She excelled at both these activities.

But Rambo, the most exceptional animal I've ever known, had other things in mind.

A couple weeks ago, I stood, incredulous, as Rambo walked up to Barbie and pawed the ground. Pawing is Rambo's signal to humans that he wants a massage—something he receives whenever he asks for it, which is generally, oh, forty or fifty times a day. Clearly he thought that if human beings could discern his wishes, then a chicken could, too. We stood there, my hand on April's arm, both of us gaping, as our extraordinary friend tried to teach his bird pal to do his bidding. When it didn't work, Rambo finally took the tip of his horn and very gently massaged the little bird. A few days later, Rambo was lying in a pile of hay. Next to him was Barbie, pulling bits of hay from his wooly coat.

The deepening of this relationship was too much for Hannah. One recent afternoon, she was nowhere to be found as I entered the barn to set up feed.

"Where's Hannah?" I asked Walt.

"She's in time out."

"What happened?" I asked, imagining his response.

"She head-butted Barbie halfway across the aisle."

Tension mounted when Barbie began using Rambo as a sofa. At seventeen pounds, she is no longer fully mobile, yet she can still manage to climb atop the resting sheep, take a moment to decide whether she wants to face his horns or his woolly rear, and ease down into fluffy bliss.

For weeks I'd heard about (but not witnessed) this new development in the relationship between the great sheep and the presumptuous chicken. And then one crisp December morning, I exited the feed room, and there, in the middle of the aisle, were Rambo and Barbie. Barbie was one happy hen plopped dead center onto Rambo's back; Rambo was completely unfazed. I dashed back inside for a camera—people *had* to see this—and then moved slowly toward them.

"Rambo, you are a prince," I praised him. "What a kind man you are," I said as I sat down just feet from them to snap the best shot.

And then I heard it: the rapid *click-click-click* of sheep hooves moving toward us. It was Hannah. She had spotted them.

The ball of brown wool pushed past me as if I weren't there and strode within six inches of the offending pair, neither of whom budged. She glared at them; she looked at me. She looked back at them; she looked at me. There was no need for words here, as *Are you going to help me here, or what?* or really, most precisely, *What the hell is this?* were etched into every gesture.

"I'm sorry, Hannah," I whispered, approaching her with consoling words. But Hannah pooped and marched outside, wanting nothing more of Rambo, the interloper, or me.

                                       ~◡〜

I never imagined I'd work at a place where a sheep and a hen would vie for a second sheep's affection. But then again, I never imagined that a dying cow would lick my face over and over and over again until he took his final breath, or that a former cock-fighting rooster would evolve into a being who begged us to share our lunches, took car rides with me, and happily climbed onto my dog's bed to share a nap.

These are the things that love allows. Animals are far more like us than I'd have ever imagined had I not had the good fortune to be with them every single day of this most wonderful life.

I hope you believe me; I hope it matters to you.

Shifting the Paradigm

Ten years ago, a fledgling CAS moved from borrowed land to a "home of our own," a derelict seventy-five-acre farm dotted with collapsed buildings, rotten fencing, and heaping piles of tires, toilets, oil tanks, appliances, and junked cars. With the help of supporters from across the country, we got to work, and in their place today are fourteen spacious pastures and twelve roomy paddocks and pens. Seven barns house our pigs, cows, goats, and sheep, while some twenty other outbuildings provide deluxe digs for our horses, donkeys, and smaller animal friends: rabbits and chickens, turkeys, ducks, and geese. Elderly and blind horses, new rescues, and various members of the "Underfoot Family," our ever-changing cast of free-range characters, occupy coveted spots in the twenty-stall, solar-powered main barn. Three houses, two office buildings, an organic garden, and a visitors' center provide for human needs.

One year ago, we purchased "CAS at 32," an abandoned property located on Rt. 32, just over two miles from us. Why an additional farm? First and foremost, Hurricane Irene taught us how urgently we needed high ground in the event of flooding.

Second, our current property is almost always filled to capacity; as long as we have other necessary resources, we want the physical space to help additional animals. Thirdly, each year CAS welcomes more visitors than in previous years. While plenty of our animals enjoy the attention, not all of them do. A second property enables us to have a true peaceful haven for that latter group, while the socialites can be housed at the more public site, where our cooking programs and special events will soon be located.

We searched for several months until a property just two miles away came on the market. We walked the property with an old map and studied its three nineteenth-century barns. To call the work of renovating the property "a project" is an understatement: hundreds of dead trees, choked by vines the size of a man's arms, must be removed. Roads must be installed; fencing erected; water lines run; barns either seriously shored up or taken down and rebuilt. Every step requires a series of approvals from the Town of Saugerties; every step involves engineers, architects, excavators, builders. *Yes:* like our original property, CAS at 32 will be both a work in progress and a labor of love for years to come. But as we expand our capacity, it is good raw material for our needs—high, rolling ground, two (possibly three) salvageable barns, one of which will be used for "people programs," and a large house that will hold a commercial kitchen, classroom space, and staff tasked with keeping a watchful eye on animal residents. We hope you'll enjoy the adventure with us! Just as importantly, we hope we can count on your support when we launch our capital campaign to fund the new farm.

Meantime, the rest of the country hasn't exactly stood still, either, in the three years since the first printing of *Animal Camp*. Truly, from where I sit, America is a dramatically different place than it was just a few short months ago. First of all, we have *finally* accepted the reality of global warming. The benefit to have come from three devastating "superstorms" and a drought that impacted a huge region of the country is that, collectively, they packed enough punch to make even the naysayers sit up and pay attention. While it remains to be seen how we respond, it's encouraging to hear policy makers having real conversations about climactic instability. We'll see if they have the courage to change food policy and food subsidies. We'll see if there's even a whisper from politicians about changing our eating habits as the truest and quickest path to reversing global warming. I suspect there won't be.

The more encouraging change is that Americans aren't waiting for politicians to step up with substantive policy initiatives: veganism has entered the mainstream in spite of corporate and political efforts to keep us sick and fat on a toxic meat and dairy-based diet. Yes, surely you've noticed, too! We're not waiting for the politicians to do the right thing. The attention veganism received in the last few months of 2012 and early 2013 alone was truly astonishing, and the statistics regarding our eating habits all the more so.

Thanks to animal rights organizations and the media, the spotlight on agribusiness makes it difficult to filter out the suffering animals endure as they are grown and killed for humans to eat them. Millions have been introduced to the concept of downed animals and to the enormity of their suffering. Millions now know that commercially grown food animals (some 99 percent of the animals grown to feed us) know nothing but suffering and terror from birth to death. In the past few years, it has become increasingly difficult to utter the words I've heard hundreds of times—"but we have laws to protect the animals!"—without sounding either stupid or something far more malevolent.

Gestation crates. Veal crates. Tens of thousands of chickens grown in a single building, sitting in their own filth, the ammonia from their excretions damaging eyes, lungs, and throats. Grown so quickly that the chicken industry now knows the percentage of birds that die of violent heart attacks because their organs can't handle the stress of being forced to grow to slaughter weight in just six weeks. We chop off pigs and cows' tails, chop off birds' toes and beaks, all without anesthesia, because on modern farms, animals packed together peck and claw and bite each other . . . and god forbid the meat be damaged. Every aspect of agribusiness is designed to maximize efficiency and profit despite the costs to the "product" (living animals) and the consumer. Americans understand this. There's more work to be done, but the information is out there, readily and frequently available, in national media and hometown newspapers, in bestselling books and award-winning documentaries, on blogs that get 100,000 hits a day and on little local radio stations. And we're paying attention.

Likewise, the impact of growing sixty-five billion animals annually to feed a human population of seven billion is wreaking environmental havoc, and we're paying attention as one disaster after the other whacks us over the head. Earth is hot and getting hotter, and climactic devastation is here. Methane from cows; the razing of the ecologically important Amazon to graze cattle or to grow what feeds them; the melting of the North Pole due to warm air being trapped by greenhouse gases; the killing of our waterways by the waste of pigs, chickens, and cows; the extinction of species— plenty of us understand that growing animals to feed people is devastatingly inefficient and even more devastatingly harmful.

Thanks to high-profile vegan doctors and the attention their work has received, we're also understanding that the consumption of animal products is killing us. Certainly the link between many forms of cancer, heart disease, stroke, obesity, type 2 diabetes, gout, high cholesterol, and other diet-related health problems have been part of the public conversation, and part of public *policy* in a few progressive pockets of the country. This is happening despite the fact that the medical establishment and the pharmaceutical industry, who make money when we're sick, might prefer it not be.

It's early yet, but the evidence of a remarkably rapid shift in our eating habits is irrefutable. A recent poll indicated that *half* of Americans are aware of the Meatless Monday campaign, and that 16 percent indicate that they eat vegetarian more than half the time. My friend Kris Carr's vegan cookbook, *Crazy Sexy Kitchen*, reached #6 on the *New York Times* bestseller list, and our local bookstore says it can't keep vegan cookbooks on the shelf. *Say what?* The USDA reports that meat consumption has dropped more than 12 percent since 2007. These are numbers that one wouldn't have dreamed possible just a few short years ago.

I'm hoping, friends, that you will help us usher in this much-needed change. I suspect that you imagined this book would be about our work to heal broken animals. In part it is. People beat and shoot and starve and abandon animals. People put them in dumpsters, or tie them to trees and move away. Catskill Animal Sanctuary will continue to welcome animals like this to our haven, where, enveloped by love, they will lead more joyful lives than they ever dreamed possible.

But here's the truth: Your own work is more important than ours will ever be. You see, the animals and the earth are pleading for help from the ones who created this mess: you and I. By eliminating animal products, each of us alone will save *thousands* of animals over the course of our lifetime. Have the courage to acknowledge your role in the suffering, *please.* To end it, to restore your health, to save the planet that sustains us all demands real solutions, not Band-Aids.

If you *really* want to help, go vegan.

All the Right Reasons

I mentioned in the prologue that there's no need for another book on agribusiness. It's true: There are many compelling, well-researched books on the subject, and I hope you'll read at least one. But if you're someone who's simply not going to do that, I hope you'll at least take ten minutes to read *All the Right Reasons,* a compilation of some of the thousands of good reasons for you to join the club!

Diet/Health

According to the American Dietetic Association, ". . . vegan diets are healthful, nutritionally adequate, and may provide health benefits in the prevention and treatment of certain diseases. Well-planned vegetarian diets are appropriate for individuals during all stages of the lifecycle, including pregnancy, lactation, infancy, child-hood, adolescence, and for athletes."

Animal products, laden with saturated fat, cholesterol, hormones, and antibiotics, are the mainstay of the typical American diet. This diet promotes heart disease, many forms of cancer, obesity, diabetes, and other chronic diseases, and is responsible for nearly a million and a half deaths annually.

More than eighty million Americans live with cardiovascular disease. The cost? Half a *trillion* dollars per year.

The good news, though, is that over 80 percent of those who switch to a healthy vegan diet reverse their cardiovascular symptoms completely and have a far lower prevalence of hypertension, the condition that causes high blood pressure.

Prostate, breast, colorectal, and pancreatic cancers are among the cancers specifically associated with eating animal products. In fact, the Physicians Committee for Responsible Medicine warns that daily meat consumption *triples* the risk of prostate and breast cancer, and regular milk consumption *doubles* the risk. Milk also increases ovarian cancer risk.

Vegetarians are 40 percent less likely to develop cancer than meat eaters.

Kidney failure, our eighth leading cause of death, may be prevented and treated with a vegan diet.

Consumption of dairy products is linked with acne and allergies.

Why do Americans get hemorrhoids? Why are we constipated? And why is there such a high incidence of colon cancer? Because we consume only twelve grams of fiber per day, as opposed to the recommended thirty-five or more. Only plant-derived foods contain fiber.

The best way to guard against osteoporosis is actually to eliminate acidifying animal products and to eat more alkalinizing fruits and vegetables and fewer acidifying animal products. Not what

Mom taught us, I know. But Mom was influenced by the National Dairy Council.

Vegans are thirty pounds lighter than meat eaters and five units lighter on the BMI (body mass index) scale.

Vegetarians and vegans are less insulin resistant than meat-eaters.

Plant-based diets help prevent, treat, and even reverse type 2 diabetes.

In one study, vegans had a 40 percent lower risk of cataracts than those eating more than 100 grams of meat per day.

Seventy percent of antibiotics—twenty-five million pounds— are fed to livestock each year. That's eight times the amount taken by humans.

The USDA's new food diagram recommends nearly half of one's diet consist of fruits and vegetables, and the meat group has been replaced by the protein group. Why is the dairy still included when it causes so many serious health issues? You guessed it: the National Dairy Council.

Around 75 percent of new human pathogens are transmissible from animals. Science discovers one new disease every four months; concentrated flocks and herds located near large human populations may be responsible.

Environment

Water

It takes between 2,500 and 5,000 gallons of water to produce a single pound of beef, versus forty-nine gallons to produce a pound of apples.

Animal feeding operations produce 500 million tons of manure every year, with factory farms generating 47–60 percent of this excrement. This is more than three times the raw waste generated by humans in the United States.

Chicken, pig, and cow poop has polluted 35,000 miles of rivers in twenty-two states and contaminated groundwater in seventeen states.

Chickens, pigs, cattle, and other animals raised for food are the primary consumers of water in the United States. A single pig consumes twenty-one gallons of drinking water per day, while a cow on a dairy farm drinks as much as fifty gallons daily.

Meanwhile, we're taking thirteen trillion gallons of water per year from the Ogallala aquifer, the largest body of fresh water on earth, mostly for beef production. As a result, say many scientists, Kansas, Oklahoma, Nebraska, Colorado, and New Mexico may soon be uninhabitable.

The Gulf of Mexico's "dead zone"—an area in which virtually all sea animals and plants have died—is half the size of Maryland. A Princeton University study found that a shift toward vegetarian eating would dramatically reduce the amount of nitrogen in the Gulf to levels that would make the dead zone "small or non-existent."

Seventy percent of potable water is used for agriculture, and due to increasing demands for meat, this is expected to increase 40 percent by 2030.

A typical CAFO manure pit is twenty-two feet deep and covers several acres. There are almost 20,000 of these "lagoons" in America. Iowa alone has had 700 manure spills in the last fifteen years. A massive North Carolina spill in 1995 dumped twenty-two million gallons into the New River and killed ten million fish.

Air

Animal agriculture is responsible for 18 percent of all human-induced greenhouse gas emissions, including 37 percent of methane emissions and 65 percent of nitrous oxide emissions. This is according to a 2006 study by the United Nations; several years later, World Watch Institute postulated that the more accurate figure was 51 percent.

The use of fossil fuels on farms to grow feed and to intensively raise land animals for food emits ninety million tons of CO_2 worldwide every year.

Globally, deforestation for animal grazing and feed crops is estimated to emit 2.4 billion tons of CO_2 every year.

In the United States, methane emissions from pig and dairy cow manure increased by 45 percent and 94 percent respectively between 1990 and 2009.

According to a study done by the Environmental Integrity Project, many factory farm test sites registered pollution emission levels well above Clean Air Act limits. Farm animals alone are expected to emit over two-thirds of the amount of greenhouse gases considered safe by 2050.

Roughly 80 percent of ammonia emissions in the United States come from animal waste.

Dairies are the largest source of the smog-making gas, surpassing trucks and passenger cars.

When the cesspools holding tons of urine and feces get full, factory farms get around water pollution limits by spraying liquid manure into the air, creating mists that are carried away by the wind.

Land

Growing corn requires more nitrogen fertilizer than any other crop, and more than half the corn in the world is fed to animals.

Raising animals for food (including land used for grazing and land used to grow feed crops) now uses nearly *one-third* of the Earth's entire land mass.

Half the world's harvest is fed to animals yet one in seven people worldwide are underfed. Growing plants to feed animals to feed people is far too inefficient to be sustainable.

Animal Cruelty

The federal Animal Welfare Act does not apply to farmed animals, and the Humane Slaughter Act specifically exempts chickens, who account for 95 percent of farmed animals. The majority of states specifically exempt "common," "customary," or "accepted" farming practices from their cruelty laws. Castration without painkiller? Not cruel. Branding? Not cruel. Cutting off tails without painkiller? Not cruel. Shoving as many chickens as can fit in a wire cage and stacking thousands of those cages together in a filthy warehouse? Egg layers spend their entire short lives in these cages, poop raining down on them from the birds stacked over their heads. As a standard agricultural practice, this is "not cruel."

A US meat inspector was labeled a "trouble maker" and was transferred to another plant in 2010 when he testified that pigs were being butchered alive where he worked. Only 1 percent of the Food Safety and Inspection Service budget is used for enforcement of the Humane Slaughter Act. USDA inspectors acknowledge that enforcement is woefully lax.

Broiler chickens raised for meat are incubated and machine hatched by the millions. Intensive single-trait breeding forces their bodies to grow to five pounds in six weeks. Those who survive this short lifespan in windowless sheds packed with tens of thousands of birds and ammonia-filled air are sent to slaughter with their baby-blue eyes and peeping like chicks (chickens don't get their final eye color until they reach sexual maturity at 14–18 weeks).

Animals on factory farms are routinely fed rendered parts of their own species, manure contaminated with hormones and antibiotics, and plastic pellets for added roughage. Yum.

Undercover video at a Smithfield facility, the world's largest pork producer, revealed sows in tiny gestation crates chewing on the bars with bloody mouths, half-dead pigs in dumpsters, and premature piglets in manure vats. Three years earlier they had pledged to phase-out such conditions.

Undercover videotape from a Pilgrim's Pride plant in West Virginia showed tearing beaks off, ripping a bird's head off, suffocating a chicken with a latex glove, and squeezing birds like balloons to spray feces on other birds.

Nine billion broiler chickens go through America's factory farms per year, suffering leg, heart, and respiratory disorders in highly concentrated, toxic environments. At the slaughterhouse they are shackled upside down by the legs and stunned enough to immobilize them, but not enough to dull the pain. Because of the speed at which the line moves, many are still alive when they get to the scalding tank and are burned to death.

The federal government uses poison, aerial gunning, steel-jawed leghold traps, and other inhumane methods to kill coyotes, mountain lions, wolves, and other "nuisance predator species" to protect livestock on private land. Other wildlife, including endangered species, are killed in the process. Even though the USDA states that most livestock losses are not predator related, $60 million is spent annually on this program.

Lest you be wooed, the term "free-range" has no legal definition in the United States. According to a report issued in 2011, 80 percent of organic eggs are produced under intense confinement in buildings housing as many as 85,000 birds with a tiny area for "outdoor access."

Other

Stop eating animals and you use dramatically less fossil fuels, as much as 250 gallons less oil per year for vegans, says Cornell University's David Pimentel, and 160 gallons less for egg-and-cheese-eating vegetarians.

According to the United Nations, a global shift toward a vegan diet is one of the steps necessary to combat the worst effects of climate change.

World meat consumption quadrupled from 1961 to 2007. The world's population is expected to increase by 30 percent by 2050 and meat production will need to double to meet consumption demands if we don't make major dietary changes. A sobering prospect for the planet.

More meat is being produced on fewer farms with prices set by a vertically integrated supply chain: Farmers and ranchers are forced into low-profit contract arrangements and animals are transformed into production units. Consumers are left with products laced with health and contamination risks and the environment around farms and slaughter plants is destroyed. No one wins.

About forty-seven billion pounds of "raw animal material"—hundreds of millions of livestock who die from stress or disease before going to slaughter, slaughterhouse scraps, road kill, and euthanized shelter animals—are sent to rendering plants each year and processed into eighteen billion pounds of soap, pharmaceuticals, personal care products, and animal feed.

If you buy chicken you're supporting economic serfdom: Contract growers are required to compete with each other and spend hundreds of thousands on capital expenditures, forcing them into perpetual debt, while companies like Tyson and Perdue control the contracts and profits.

References

"2011 World Hunger and Poverty Facts and Statistics." worldhunger.org.

Adebamowo, C. A., D. Spiegelman, F. W. Danby, et al. "High School Dietary Dairy Intake and Teenage Acne." *Journal of the American Academy of Dermatology* 52, no. 2 (2005): 207–14.

Adventist Health Study. Loma Linda University. http://www.llu.edu/public-health/health/lifestyle_disease.page.

Aillery, Marcel, Noel Gollehon, Robert Johansson, Johnathan Kaplan, Nigel Key, Marc Ribaudo. "Managing Manure to Improve Air and Water Quality: A Report from the Economic Research Service." *Economic Research Report 9* (2005), last accessed June 17, 2008. www.ers.usda.gov/publications/ERR9.

The American Dietetic Association. "Position of the American Dietetic Association: Vegetarian Diets. *Journal of the American Dietetic Association* 109, no. 7 (2009).

The American Heart Association. "Fiber," last accessed April 11, 2006. www.americanheart.org/presenter.jhtml?identifier=4574.

The American Heart Association. "Heart Disease and Stroke Statistics—2010 Update: A Report from the American Heart Association." *Circulation* (2010).

The American Heart Association. "Heart Disease and Stroke Statistics 2010" (2010).

Anon. "Dietary Fiber and Colon Cancer: the Pendulum Swings (Again)." *Harvard Men's Health Watch* 10, no. 1 (2005): 1–5.

Appleby, P. N., N. E. Allen, T. J. Key. "Diet, Vegetarianism, and Cataract Risk." *American Journal of Clinical Nutrition* 93, no. 5 (2011): 1128–35.

Bair, Julene. "Running Dry on the Great Plains." *New York Times*, Nov 30, 2011.

Barkema, Alan, Mark Drabenstott, Nancy Novack. "The U.S. Meat Industry." *Main Street*, April 2001.

Becker, Geoffrey S. "Animal Rendering: Economics and Policy." *Congressional Research Service Reports* (March 17, 2004).

Bittman, Mark. "Rethinking the Meat-Guzzler." *New York Times*, January 27, 2008.

Bottemiller, Helena. "USDA Vet Blows Whistle on Food Safety Agency." *Food Safety News*, March 10, 2010.

Boyle, Lisa Kaas. "Did You Just Eat a Plastic Bag?" *Huffington Post*, January 6, 2011.

Brody, Jane. "Exploring a Low-Acid Diet for Bone Health." *New York Times*, November 23, 2009.

Butler, Virgil. "Clarification on Stunner Usage." *United Poultry Concerns* (May 27, 2004). http://www.upc-online.org/slaughter/91104stunner.htm.

Center for Disease Control. www.cdc.gov.

Center for Food Security and Public Health. "Animal Disease List." http://www.cfsph.iastate.edu/DiseaseInfo.

Compassion Over Killing. www.cok.net.

Cornucopia Institute Organic Egg Report and Score Card. www.cornucopia.org/organic-egg-scorecard/.

Craig, Winston J. "Health Effects of Vegan Diets." *American Journal of Clinical Nutrition* (2009).

Danby, F. William, MD. "Nutrition and Acne." *Clinics in Dermatology* 28, no. 6 (2010): 598–604.

Doering, Christopher. "Manure Causes Stink for Lawmakers and Farmers" *Reuters*, September 6, 2007.

Donner, Simon D. "Surf or Turf: A Shift from Feed to Food Cultivation Could Reduce Nutrient Flux to the Gulf of Mexico," *Global Environmental Change* 17 (2007): 105–13.

Doorn, Michiel R. J., David F. Natschke, Pieter C. Meeuwissen. *Review of Emission Factors and Methodologies to Estimate Ammonia Emissions from Animal Waste Handling.* North Carolina: National Risk Management Research Laboratory, 2002.

Dr. Barnard's Blog; "Power to the Plate: Food Subsidies Should Reflect New USDA Dietary Advice," blog entry by Neal Barnard, June 2, 2011.

Environmental Integrity Project. "Hazardous Pollution from Factory Farms: An Analysis of EPA's National Air Emissions Monitoring Study Data" (2011).

Environmental Protection Agency. "Compliance and Enforcement National Priority: Concetrated Animal Feeding Operations (CAFOs)" (2009).

Farm Sanctuary. www.farmsanctuary.org.

Fitzenberger, Jennifer M. "Dairies Gear UP for Fight Over Air." *Fresno Bee*, August 2, 2005.

Franco, A., A. K. Sikalidis, J. A. Solís Herruzo. "Colorectal Cancer: Influence of Diet and Lifestyle Factors." *Revisto Española de Enfermedades Degistivas* 97, no. 6 (2005): 432–48.

Goodland, Robert, Jeff Anhang. "Livestock and Climate Change: What if the key actors in climate change are . . . cows, pigs and chickens?" *World Watch* (2009).

Goodner, David. "Manure-on-Snow Ban Begins Tomorrow." IowaPolitics.com, December 20, 2010.

Ho-Pham, Lan T., Nguyen D. Nguyen, Tuan V. Nguyen. "Effect of Vegetarian Diets on Bone Mineral Density: a Bayesian Meta-Analysis." *The American Journal of Clinical Nutrition* (2009).

Ho-Pham, L. T., P. L. Nguyen, T. T. Le, T. A. Doan, N. T. Tran, T. A. Le, T. V. Nguyen. "Veganism, Bone Mineral Density, and Body Composition: a Study in Buddhist Nuns." *Osteoporosis International* (2009).

Hoekstra, Arjen. "The Water Footprint of Food," in *Water for Food*, edited by J. Förage, 49–60. Stockholm, Sweden: The Swedish Research Council for Environment, Agricultural Sciences and Spatial Planning, 2008.

"Home is Where the Germ Is: Keeping Bugs at Bay in the Kitchen." *Nutrition Action Healthletter*, January 1, 2004.

"Hot Spots: How Changing Farm Habits Threaten Public Health." *The Economist*, February 10, 2011.

"HSUS Exposes Inhumane Treatment of Pigs at Smithfield" (December 15, 2010). www.humanesociety.org/news/press_releases/2010/12/smithfield_pigs_121510.html.

"Humane Methods of Slaughter Act, Weaknesses in USDA Enforcement: Testimony before the Subcommittee on Domestic Policy, Committee on Oversight and Reform, House of Representatives" (March 4, 2010)

"Interview with Dr. Esselstyn and Dr. Ornish." The Situation Room with Wolf Blitzer (October 4, 2010).

Inventory of U.S. Greenhouse Gas Emissions and Sinks: 1990–2008. U.S. Environmental Protection Agency, 2010.

"Is Agriculture Depleting Our Water Supply on Purpose?" last accessed December 9, 2010. Investmentu.com.

King, Lonnie, DVM. "Zoonotic Diseases, Part 1." By Teddi Dineley Johnson. American Public Health Association, Get Ready campaign (2008). www.getreadyforflu.org.

Lee, Jennifer. "Neighbors of Vast Hog Farms Say Foul Air Endangers Their Health." *New York Times*, May 11, 2003.

Leutwyler, Kristin. "Most U.S. Antiobotics Fed to Healthy Livestock." *Scientific American*, January 2001.

Lin, J. F. B. Hu, G. C. Curhan. "Associations of Diet with Albuminuria and Kidney Functions Decline." *Clinical Journal of the American Society of Nephrology* (May 2010).

Markarian, Michael. "Congress Can Spare Taxpayers and Animals." *Huffington Post*, December 7, 2010.

Mathur, R. V., J. R. Shortland, A. M. el-Nahas. "Calciphylaxix." *Postgrad Medical Journal* (September 2001).

McNeil, Donald G., Jr. "KFC Supplier Accused of Animal Cruelty." *New York Times*, July 20, 2004.

Mekonnen, Mesfin, Arjen Hoekstra. "A Global Assessment of the Water Footprint of Farm Animal Products." *Ecosystems* 15, no. 3 (2012): 401–15.

Moe, S. M., M. P. Zidehsarai, M. A. Chambers, L. A. Jackman, J. S. Radcliffe, L. L. Trevino, S. E. Donahue, J. R. Asplin. "Vegetarian Compared with Meat Dietary Protein Source and Phosphorous Homeostasis in Chronic Kidney Disease. *Clinical Journal of the American Society of Nephrology* (February 2011).

Monbiot, George. "Why Vegans Were Right All Along." *The Guardian*, December 24, 2002.

National Heart, Lung, and Blood Instititue. *Fiscal Year Fact Book 2009.*

National Sustainable Agriculture Coalition. sustainableagriculture.net/our-work/conservation-environment/clean-water-act.

Ornish, Dean, MD. "Mostly Plants." *American Journal of Cardiology* 104, no. 7 (2009): 957–58.

Pan, Min-Hsiung, Ching-Shu Lai, Slavik Dushenkov, Chi-Tang Ho. "Modulation of Inflammatory Genes by Natural Dietary Bioactive Compounds." *Journal of Agricultural and Food Chemistry* 57, no. 11 (2009): 4467–77.

Pelletier, Nathan, Peter Tyedmers. "Forecasting Potential Global Environmental Costs of Livestock Production 2000–2005." *Proceedings of the National Academy of Sciences of the United States of America* (2010).

The Pew Enviromental Group. "Antitrust Law, Corporate Concentration and the Meat-Processing Industry" (2010).

Physicians Committee for Responsible Medicine. "Meat Consumption and Cancer Risk."

http://pcrm.org/health/cancer-resources/diet-cancer/facts/meat-consumption-and-cancer-risk.

Pimental, David, Marcia Pimentel, "Sustainability of Meat-Based and Plant-Based Diets and the Environment." American Journal of Clinical Nutrition 78, no. 3 (2003).

Prentice Ann. "Diet, Nutrition and the Prevention of Osteoporosis." *Public Health Nutrition* 7, no. 1A (2004): 227–43.

Ray, Daryll, Hardwood D. Schaffer. "Policy Pennings." *MidAmerica Farmer Grower* 31, no. 14 (2011).

"Rendering Plants: Recycling of Dead Animals and Slaughterhouse Wastes." *Muslim Observer*, June 23, 2007.

Rice, Pamela. *101 Reasons Why I'm Vegetarian*. New York: Lantern Books, 2005.

Sarter, B., T. C. Campbell, J. Fuhrman. "Effect of a High Nutrient Density Diet on Long-Term Weight Loss: a Retrospective Chart View." *Alternative Therapies in Health and Medicine* 14, no. 3 (2008): 48–53.

Shapiro, Paul. "Factory Farming: A Report Card - Part 2 of 5" (2009). YouTube.com.

Shlachter, Barry. "Contract Growers Hoping the Chicken Industry Offers a Steady Nest Egg; May Instead be Trapped by Debt." *Star-Telegram*, March 6, 2005.

Spencer, E. H., H. R. Ferdowsian, N. D. Barnard. "Diet and Acne: A Review of the Evidence." *International Journal of Dermatology* 48, no. 4 (2009): 339–47.

Steinfeld, H., P. Gerber, T. Wassenaar, V. Castel, M. Rosales, C. de Haan. *Livestock's Long Shadow: Enviromental Issues and Options*. Rome: Food and Agriculture Organization of the United Nations, 2006.

Sunstein, Cass R., Martha C. Nussbaum. *Animal Rights: Current Debates and New Directions*. New York: Oxford University Press, 2005.

Takeda, E., H. Yamamoto, Y. Nishida, T. Stat, N. Sawada, Y. Taketani. "Phosphate Restriction in Diet Theraphy." *Contributions in Nephrology* (2007).

Tonstad, S., T. Butler, R. Yan, G. E. Fraser. "Type of Vegetarian Diet, Body Weight, and Prevalence of Type 2 Diabetes." *Diabetes Care* 32, no. 5 (2009): 791–96.

Union of Concerned Scientists. "They Eat What?: The Reality of Feed at Animal Sanctuaries" (August 8, 2006). www.ucsusa.org/food_and_agriculture/our-failing-food-system/industrial-agriculture/they-eat-what-the-reality-of.html.

United Poultry Concerns (UPC). www.upc-online.org.

"USDA Unveils MyPlate as the New Food Icon." *International Business Times*, June 13, 2011.

"Vegan Diet Reduces Breast Cancer Risk." The Cancer Project, last accessed May 2011. support.cancerproject.org/site/MessageViewer?em_id=12001.0.

Wetzler, Andrew. "Wildlife Services: The Most Important Wildlife Agency You've Never Heard Of." *Huffington Post*, Dec 4, 2009.

Wiwanitkit, Viroj. "Renal Function Parameters of Thai Vegans Compared with Non-Vegans." *Renal Failure* (2007).

Your Health: Vegetarian Foods: Powerful for Health," last accessed June 2011. http://www.pcrm.org/health/diets/vsk/vegetarian-starter-kit-powerful.

APPENDIX TWO

Suggested Reading, Eating, Viewing

Books

- Balcombe, Jonathan. *Pleasurable Kingdom* (New York: Macmillan, 2007). Balcombe, a leading animal behavior researcher, provides rigorous evidence that all creatures experience the same positive feelings as humans with regard to play, sex, touch, food, anticipation, comfort, and more.
- Barnard, Neal. *Eat Right, Live Longer* (New York: Three Rivers Press, 1995). In an eight-step program, learn how specific vegetarian food choices help you to prevent diseases, live longer, and experience more energy. Recipes and menus are included.
- Bekoff, Marc. *The Emotional Lives of Animals: A Leading Scientist Explores Animal Joy, Sorrow, and Empathy—and Why They Matter* (Novato, California: New World Library, 2007). Through years of studying social communication of a wide range of animal species, this scientist's touching stories confirm the existence of emotions including joy, empathy, grief, anger, and love.

- Campbell, T. Colin and Thomas M. Campbell, II. *The China Study: The Most Comprehensive Study of Nutrition Ever Conducted and the Startling Implications for Diet, Weight Loss, and Long-Term Health* (Dallas, Texas: BenBella Books, 2006). Based on the most comprehensive large study ever conducted, Dr. Campbell presents the powerful connection between what you eat and the impact on your health and well-being.

- Conkin, Paul. *Revolution Down on the Farm: The Transformation of American Agriculture Since 1929* (Lexington, Kentucky: The University Press of Kentucky, 2008). The author documents agricultural changes over the past eighty years from the small farms of the early twentieth century to the current farm factories.

- Dawn, Karen. *Thanking the Monkey: Rethinking the Way We Treat Animals* (New York: Harper Collins, 2008). A light and delightful read with a serious message about how we treat animals.

- Eisman, George. *The Most Noble Diet* (Diet Ethics, 1994). How your diet affects you and the world. Food guides and recipes provided.

- Eisnitz, Gail. *Slaughterhouse: The Shocking Story of Greed, Neglect, and Inhumane Treatment Inside the U. S. Meat Industry* (Amherst, New York: Prometheus Books, 2006). The author, chief investigator for the Humane Farming Association, exposes the corruption and cruelty of slaughterhouses and the ongoing efforts to improve conditions in the meatpacking industry.

- Esselstyn, Caldwell, Jr. *Prevent and Reverse Heart Disease* (New York: Avery, 2007). An internationally known surgeon, researcher, and former Cleveland Clinic clinician, Dr. Esselstyn illustrates how to prevent and help reverse the effects of heart disease by adopting a plant-based diet.

- Fox, Michael. *Eating with Conscience: The Bioethics of Food* (Troutdale, Oregon: NewSage Press, 1997). Hard-hitting eye-opener written by a vet.

- Hatkoff, Amy. *The Inner World of Farm Animals: Their Amazing Social, Emotional, and Intellectual Capacities* (New York: Stewart, Tabori & Chang, 2009). A beautiful book that encourages the reader to understand the individuality of each single farm animal.
- Hauter, Wanonah. *Foodopoly: The Battle Over the Future of Food and Farming in America* (New York: The New Press, 2012). An exhaustively researched book that examines corporate control of our food supply. A frightening but important read.
- Joy, Melanie. *Why We Love Dogs, Eat Pigs, and Wear Cows: An Introduction to Carnism* (San Francisco, California: Conari Press, 2011). We all like to think we form our own opinions but this book challenges that assumption by examining how we have been conditioned to view certain animals as more or less deserving of human compassion.
- Lyman, Howard. *Mad Cowboy: Plain Truth from the Cattle Rancher Who Won't Eat Meat* (New York: Scribner, 1998). The author's statements about the "livestock industry" led Oprah Winfrey to declare she'd never eat another hamburger. Find out why.
- Lyman, Howard with Glen Merzer and Joann Samorow-Merzer. *No More Bull! The Mad Cowboy Targets America's Worst Enemy: Our Diet* (New York: Scribner, 2005). So you think the FDA is keeping our food supply safe? Not so fast!
- Masson, Jeffrey Moussaieff. *The Face on Your Plate: The Truth About Food* (New York: W. W. Norton & Company, 2009). The planet and every creature who inhabits it are inexorably linked. What we do to them we also do to ourselves.
- McMillan, Tracie. *The American Way of Eating: Undercover at Walmart, Applebee's, Farm Fields, and the Dinner Table* (New York: Scribner, 2012). What if you can't afford $4 for an organic tomato but still want to eat good healthy food? An engrossing and thoughtfully written book.

- Nestle, Marion. *Food Politics: How the Food Industry Influences Nutrition and Health* (Berkeley, California: University of California Press, 2007). What happens when food production exceeds food consumption? Simple: producers get the public to consume more! This book shows you how they influence Congress and change the eating habits of a nation.
- Regan, Tom. *Empty Cages: Facing the Challenge of Animal Rights* (Lanham, Maryland: Rowman and Littlefield, 2004). This book has been described as one of the best introductions to the issue of animal rights and makes for compelling reading.
- Rifkin, Jeremy. *Beyond Beef* (New York: Penguin Books, 1992). The cost of a hamburger far exceeds the few bucks we hand over at McDonald's. This book spells out the true cost with unflinching honesty.
- Robbins, John. *Diet for a New America: How Your Food Choices Affect Your Health, Happiness and the Future of Life on Earth, 2nd edition* (Tiburon, California: HJ Kramer/New World Library, 2012). From the iconic heir to the Baskin Robbins fortune, this iconic book raises compelling questions that are impossible for any thinking person to ignore.
- Robbins, John. *May All Be Fed: A Diet for a New World* (New York: Harper Perennial, 1993). A compelling examination of the inequities of food distribution and how our personal choices affect world hunger.
- Rowe, Martha. *The Way of Compassion* (Herndon, Virginia: Lantern Books, 2000). How are animal rights, environmentalism, social justice, and your diet connected? Shared compassion and the quest for a world free from exploitation.
- Safran Foer, Johnathan. *Eating Animals* (New York: Back Bay Books, 2010). A father's philosophical journey to answer the question, "Why do we eat some animals and not others?"

- Schlosser, Eric. *Fast Food Nation: The Dark Side of the All-American Meal* (New York: Houghton Mifflin, 2001). Fast food permeates American culture, but few people consider the devastating consequences that this fascination causes.
- Scully, Matthew. *Dominion: The Power of Man, the Suffering of Animals, and the Call to Mercy* (New York: St. Martin's Press, 2002). A powerful look at the hypocrisy of the Bible's message to respect life and the ways that mankind excuses systemic abuse.
- Singer, Peter (editor). *In Defense of Animals: The Second Wave* (Malden, Massachusetts: Wiley-Blackwell, 2006). A collection of essays by philosophers and activists in the animal rights movement.
- Singer, Peter and Jim Mason. *The Way We Eat: Why Our Food Choices Matter* (Emmaus, Pennsylvania: Rodale Books, 2007). Lays out the "five principles for making conscientious food choices," and follows three American families with very different dietary habits to track the sources of their food and what ethical issues they should learn about.
- Stepaniak, Joanne. *Being Vegan: Living with Conscience, Conviction, and Compassion* (Lincolnwood, Illinois: Lowell House, 2000). A Q&A on being vegan in all aspects of your life.
- Tuttle, Will. *The World Peace Diet* (Herndon, Virginia: Lantern Books, 2005). Reconnect to the natural and spiritual world by examining your relationship to the plants, animals, and lifecycle of what you eat. Tuttle argues persuasively that the root of all violence in the world is our consumption of animals.

Blogs

Dr. Barnard's Blog, www.pcrm.org/media/blog. Dr. Neal Barnard, MD and founder of Physicians Committee for Responsible Medicine and the Cancer Project.

Crazy Sexy Wellness, www.kriscarr.com/blog. A weekly serving of love and wisdom that will change your life.

Girlie Girl Army: Your Glamazon Guide to Green Living, www. girliegirlarmy.com.

Healthy Happy Life, www.kblog.lunchboxbunch.com. Vegan recipes, photos, and wellness tips for a healthy happy life.

The Discerning Brute, www.thediscerningbrute.com. Joshua Katcher: food, fashion, and etiquette for the ethically handsome man.

Meatless Monday, www.meatlessmonday.com. Meatless Monday is a non-profit initiative of the Monday Campaigns, in association with the Johns Hopkins' Bloomberg School of Public Health. They provide the information and recipes you need to start each week with healthy, environmentally friendly meat-free alternatives.

Our Hen House, www.ourhenhouse.org. A multi-media hive of opportunities for change whose mission is to effectively mainstream the movement to end the exploitation of animals.

This Dish is Veg, www.thisdishisvegetarian.com. Dishing up animal rights, vegan, vegetarian, and eco-friendly news.

Vegan Mainstream, www.veganmainstream.com/veganblogs. Vegan viewpoints on a variety of current events and trends.

Vegans of Color, www.vegansofcolor.wordpress.com All oppressions are connected.

Your Daily Vegan, www.yourdailyvegan.com. To give a voice to the people who are steadfast and unapologetic about the rights and autonomy of non-humans.

Suggested Eating

You'll be amazed by the number of vegan cookbooks on the market. Here are just a few of my favorites.

Asbell, Robin. *Big Vegan: More than 350 Recipes, No Meat/No Dairy All Delicious* (San Francisco, California: Chronicle Books, 2011). Whether you're a veggie newbie or a long-time vegan, this book has easy, satisfying recipes you'll want to try for all three meals of the day.

Atlas, Nava. *Vegan Soups and Hearty Stews for All Seasons* (New York: Clarkson Potter, 2009). Another one of our favorites from Nava Atlas. Soups from around the world, hot or cold, chunky or smooth, even dessert soup!

Atlas, Nava. *Wild About Greens: 125 Delectable Vegan Recipes for Kale, Collards, Arugula, Bok Choy, and Other Leafy Veggies Everyone Loves* (New York: Sterling, 2012). Amazing recipes for seventeen types of leafy greens—even dedicated foodies will find something new to try.

Atlas, Nava and Susan Voisin. *Vegan Holiday Kitchen: More than 200 Delicious, Festive Recipes for Special Occasions* (New York: Sterling 2011). The name really says it all! All the food you need to celebrate Thanksgiving, Christmas, Chanukah, July 4th, your birthday, and every other special occasion.

Brazier, Brendan. *Thrive Foods: 200 Plant-Based Recipes for Peak Health* (Cambridge, Massachusetts: Da Capo Lifelong Books, 2011). Athletes and anyone else wanting to up their game by eating plant super foods needs this book. Even the table of contents will make your mouth water! Competitive athletes should check out the entire *Thrive* series.

Carr, Kris. *Crazy Sexy Diet: Eat Your Veggies, Ignite Your Spark, and Live Like You Mean It!* (Guilford, Connecticut: Skirt!, 2011). For everyone looking for the motivation to eat healthy, sprinkled with recipes to get you started.

Carr, Kris. *Crazy Sexy Kitchen: 150 Plant-Empowered Recipes to Ignite a Mouthwatering Revolution* (Carlsbad, California: Hay House, 2012). Kris shows you that healthy eating and deprivation do *not* go hand in hand! Smoothies, salads, and, yes, dessert!

Freston, Kathy. *Veganist: Lose Weight, Get Healthy, Change the World* (New York: Weinstein Books, 2009). If you're a junk food junky who needs a gradual step-by-step way to change your eating habits, then this one's for you!

Freedman, Rory and Kim Barnouin. *Skinny Bastard: A Kick-in-the-Ass for Real Men Who Want to Stop Being Fat and Start Getting Buff* (Philadelphia, Pennsylvania: Running Press, 2009). Eating veggies won't make you a girlie girl. This is the guy's guide to going veg.

Freedman, Rory and Kim Barnouin. *Skinny Bitch* (Philadelphia, Pennsylvania: Running Press, 2005). When you're ready for the truth about "dieting," with a little attitude!

Hester, Kathy. *The Vegan Slow Cooker: Simply Set It and Go with 150 Recipes for Intensely Flavorful, Fuss-Free Fare Everyone (Vegan or Not!) Will Devour* (Beverly, Massachusetts: Fair Winds Press, 2011). Wouldn't you love to have a hearty warm meal waiting for you when you get home? We thought so!

Jones, Ellen Jaffe. *Eat Vegan on $4.00 a Day: A Game Plan for the Budget Conscious Cook* (Summertown, Tennessee: Book Publishing Company, 2011). Vegan doesn't have to be expensive. Stock up, plan your meals, and save a bundle.

Moscowitz, Isa Chandra. *Vegan with a Vengeance: Over 150 Delicious, Cheap, Animal-Free Recipes That Rock* (Cambridge, Massachusetts: Da Capo Press, 2005). Homestyle meals with a little attitude and a good sense of humor.

Moscowitz, Isa Chandra. *Vegan Brunch: Homestyle Recipes Worth Waking Up for—from Asparagus Omelets to Pumpkin Pancakes* (Cambridge, Massachusetts: Da Capo Lifelong Books, 2006). Soon you'll be hosting brunches just to show off your delicious creations!

Moscowitz, Isa Chandra and Terry Hope Romero. *Veganomicon: The Ultimate Vegan Cookbook* (Cambridge, Massachusetts: Da Capo Lifelong Books, 2007). Any cookbook by Isa or Terry will get lots of use, but if can only choose one, this one has a little of everything and covers all the bases.

Noyes, Tamasin and Celine Steen. *Vegan Sandwiches Save the Day!: Revolutionary New Takes on Everyone's Favorite Anytime Meal* (Beverly, Massachusetts: Fair Winds Press, 2012). Never again wonder what to pack for lunch.

Patrick-Goudreau, Colleen. *The 30-Day Vegan Challenge: The Ultimate Guide to Eating Cleaner, Getting Leaner, and Living Compassionately* (New York: Ballantine, 2011). Get animal products out of your kitchen one day at a time.

Patrick-Goudreau, Colleen. *Vegan's Daily Companion: 365 Days of Inspiration for Cooking, Eating, and Living Compassionately* (Beverly, Massachusetts: Quarry Books, 2011). When you're done with the thirty-day challenge, get inspiration and recipes for staying vegan all year. Features an essay by Kathy Stevens and our very own Rambo is on the front cover.

Pierson, Joy, Angel Ramos, and Jorge Pineda. *Candle 79 Cookbook: Modern Vegan Classics from New York's Premier Sustainable Restaurant* (Emeryville, California: Ten Speed Press, 2011). Sometimes you want to cook fancy, and this book will show you how.

Robertson, Robin. *1,000 Vegan Recipes* (Hoboken, New Jersy: John Wiley & Sons, 2009). Something for everyone. Literally.

Romero, Terry Hope. *Vegan Eats World: 300 International Recipes for Savoring the Planet* (Cambridge, Massachusetts: Da Capo Lifelong Books, 2012). Around the world in eighty plates . . . wait, make that 300 plates!

Schinner, Miyoko. *Artisan Vegan Cheese* (Summertown, Tennessee: Book Publishing Company, 2012). Much easier than it sounds! If you thought you could never go vegan because you'd have to give up cheese, think again.

Sroufe, Del. *Forks Over Knives – The Cookbook: Over 300 Recipes for Plant-Based Eating All Through the Year* (New York: The Experiment, 2012). Great for everyday meals, with contributions by some of our favorite authors.

Terry, Bryant. *The Inspired Vegan: Seasonal Ingredients, Creative Recipes, Mouthwatering Menus* (Cambridge, Massachusetts: Da Capo Life-long Books, 2012). Healthy takes on some Southern favorites, with an international twist.

Suggested Viewing

Earthlings (Shaun Monson/Nation Earth, 2005)

Farm to Fridge (Lee Iorvino/Mercy for Animals, 2011)

Fast Food Nation: The Dark Side of the All-American Meal (Richard Linklater/Fox Searchlight, 2006)

Fat, Sick, and Nearly Dead (Joe Cross/Reboot Media, 2011)

Food, Inc. (Robert Kenner/Magnolia Pictures, 2008)

Forks Over Knives (Lee Fullerson/Monica Beach Media, 2011)

King Corn (Aaron Woolf/New Video Group, 2008)

May I Be Frank? (Ryan Engelhart et. al./Cinema Libre Studio, 2013)

Meet Your Meat (Bruce Friedrich/PETA, 2002)

Peaceable Kingdom: The Journey Home (Jenny Stein/Tribe of Heart, 2012)

Super Size Me (Morgan Spurlock/Sony Pictures, 2004)

Vegucated (Marisa Miller Wolfson/Filmbuff, 2011)

The Witness (Jenny Stein/Tribe of Heart, 2004)

Acknowledgments

Everything I needed to know about what matters in life, I learned from my grandmother, Eleanor Ann Pickup Furman. "Grannylou" was the finest human being I've ever known. I am grateful beyond measure for her life and her lessons.

Dad, you taught me so many of the right things. You still do: Do what you love, and do it with your whole heart. Be brave. Embrace life. Laugh a lot. Offer joy. Stay in touch. Thanks, Pop.

My sister Ellen, brother-in-law Jay, and their brood—Sara, Brandon, Courtney, Riley—are living proof of the power of love. A truly remarkable family, they ground me more than they can possibly know. So does my precious brother Ned . . . when he remembers to call. And to the entire Nashville clan, especially my dear aunt Beverly Ann, great aunt Beverly, and my cousin, Jill St. John: Thank you for the light you shine. Beverly Ann: you're my hero. You were Grannylou's, too.

Without Jesse Moore, there would have been no Catskill Animal Sanctuary. Thanks, good man. Godspeed on your new adventure.

How grateful I am for the love, determination, and team spirit with which the current CAS crew greets each day, no matter its challenges. My heartfelt thanks to: Michelle Alvarez, Julie Barone, Leah Craig Chumbley, Caleb Fieser, Kathy Keefe, Jenn Mackey, Russ Mackey, Rebecca Moore, Erin Murphy, Linda Soper-Kolton, Alex Spaey, and Robyn Welty. Gang, you make dreams possible!

Whether for one year or eight, some good people contributed enormously in service to something larger than themselves. I am especially grateful to former staff members Kevin Archer, Walt Batycki, Troy Gangle, April Harrison, Allen Landes, Abbie Rogers, Lorraine Roscino, and Karen Wilson.

Betsy Messenger and Melissa Bamford will be tough acts to follow as we recruit new teachers for our fourth season of Camp Kindness! Thanks to both of them for shaping a program that changes children's lives, and my sincere apologies that there's no Camp Kindness chapter to highlight your life-changing work. It is a glaring omission. Thanks to Melissa for all that she brings so generously to our community.

David Cooper: you are my best friend. You were Hannah's best friend, too—the best daddy a dog could have asked for. Thank you for doing the heavy lifting: She was the happier dog for her life with you.

Like my dear Grannylou, my dear neighbor Frank Tiano, Sr. reminds me of what matters: Slow down. Help people. Smile. Thank you to Frank and Kathy for tolerating with such grace all the goings on of their neighbors down the hill.

To the good folks at Skyhorse: Thanks for allowing me to breathe new life into this book. Get ready: There are more to come! Thanks to Nicole Frail for giving me "one more week."

Tanya Thomas: Thank you for coming along when you did, and for your love and clarity in helping me with my do over.

How on Earth do we do what we do with such a lean staff? Answer: VOLUNTEERS! At any given time, we have forty or more active volunteers—folks who give a minimum of a half-day per week. A heartfelt thanks to those who have shared their time and/or talents with us, and a plea for forgiveness to the hundreds left off the list due to space constraints. Please know that we deeply appreciate all that you've done for the critters who call CAS home, and for voiceless farm animals everywhere. A special thanks to some of the "heavy lifters" below:

Debbie Allen, Dorothy Avery, John and Amenda Baker, JC Barone, Jamie Becker, Andrew Berthiame, Cathy Blake, Teddy Blake, Rich Bollin, Paul Bouros, Elena Brandhofer, Louise Brinkerhoff, Lyn Brown, Juli Buono, Christine "Veggie Burger," Don and Denise Carson, Melanie Carpenter, Ani Castillo, Barbara Chapman, Sanjina Choudhury, Nancy Clark, Tami Colwell, Alex Cooper, Doris Coursen, Eileen Cunningham, Lisa Cutten, Julie and Norman Desch, Dave DiNicola, Julianne Dow, Shawn and Billy Dougherty, Emmah Donohue, Sarah Draney, Marion DuBois, Jan Durand, Dianna Ficker, Jacquelyn Fishburne, Aaron Flaherty, Maureen Ford, Dawn Freedman, Joanne Friedman, Maureen Ford, Lindsey Gavette, Judi Gelardi, Mike Graff, Deanna Gray, Abby and Rudy Hacker, Caryn Haggerty, Mary Heyer, Dawn Dixon Hubbell, Beverly Harris, Julie Harris, Sunil Joseph, Elena Kastenbaum, Gary Kaiser, Julie Kirkpatrick, Eli and Tammy Kassirer, Natalie Korniloff, Deborah LaFond, Marvin Lang, Stefanie Lang, Donna Lenhart, Helen Levine, Karen Lockrow, Elaine Luczka, Karen Markisenis, Jan Marotta, Pascale Martel, Christine Martin, Walter McGrath, Mary Ellen Meitus, Charlotte Mollo, Mary Ellen Moore, Kisha Munson, Jon Novick, Brian Normoyle, Eleanor Olds, Paula Pell, Colette Perfitt, Betty Osterhoudt, Andrea Plotkin, Nancy Purdum, Donna Reynolds, Susan Rich, Chris Rosenburg, the Rubsam

Family, especially Siobhan, Anne Rugh, Judy Samoff, Jane Sarcona, Ellie Sarty, Carol Sas, Caleb Schneiderman, Bonnie Schweppe, John Schoonmaker, Chris Seeholzer, Roni Shapiro, Carol Smith, Andrea Shaut, Connie Snyder, Kathy Somma, Kim Splain, Courtney and Riley Sullivan, Kylie Standish, Clara Steinzor, Violet Streich, Pat Thurston, Amanda Tiffany, Kelly Tomaseski, Janet Treadaway, Dara Trahan, Christine Trembaly, Alisha Utter, Vanessa Van Noy, Samaritan Village, Jackie Vilmany, David Voight, Lindsey Weidman, David Wemple, Deb Weir, Sean Welty, Barbara Wood, Anthony Zitelli, and Maureen Zoellner.

To adequately depict the dedication and generosity of Donna Albright, Phyllis Kaiser, Jill Meyers, Lois Samsel-Cronk, and Julie Buono would take an entire chapter. They are among the most giving people I have ever known. It's actually fair to say that without Donna's and Julie's tireless efforts in the barn, we'd need another staff person. Phyllis and Lois, meanwhile, do quadruple duty, always with a smile: They work in the barn, the office, the Welcome Hut, and the vegan kitchen! And Jill scoops poop, organizes auctions, and donates exquisite photographs, all with a smile. We love you, ladies.

To our donors, large and small: Thank you for being a part of this challenging, vital work. Without you, there would be no Catskill Animal Sanctuary.

My deepest gratitude to the professionals, colleagues, and plain old friends who've shared either time, labor, talent, perspective or guidance in the interest of the animals: Sara Alexander, Joyce Anderson, ASPCA, Jonathan Balcombe, Susan Barnett, Andrea Barrist-Stern, Jennifer Brown Consulting, Judy Boruta, Harold

Brown, Kris Carr, Holly Cheever, Cathy Cloutier, Susie Coston, Dick Crenson, David DiNicola, George Eisman, Len Egert, Flickr Filmworks, Greg Gattine, Laurie Goldstein, David Green, Healthy Gourmet to Go, Corey Hedderman, Hudson Valley Compassion, Hurley Veterinary Hospital, Jivamukti Yoga, Karma Road, David Life, Ruth Lipman, Sue McDonough, Carol Meyer, Mid-Hudson Vegetarian Society, Montclair Vegan Society, Veggie Conquest, Mother Earth's Storehouse, Dr. Heather O'Leary, Dr. Gary Patronek, Mark and Phil, Kevin Post, Jen Redmond, Kathy Ruttenberg, Jen Sauer, David Sax, Jill Shufeldt, The Seed, Colleen Segarra, Chris Seeholzer, Brian Shapiro, Paul Shapiro, Dr. Andrea Sotela, Bill Spearman, Jill Spero, Tim Sweeney, Frank Tiano, Jr., Amy Trakinski, Judith Turkel, Wendy Van Aken, Bill Wolfsthal.

Finally, to the animal friends no longer with us, especially Samson, Rambo, Murphy, Babe, Paulie, and Hannah: Thanks for showing me the way.

About the Photographers

Photographer **David Sax** grew up north of Chicago and has lived in many beautiful regions of the country, including Colorado, Washington, Arizona, and upstate New York. In 2004, he left his work as a physical therapist specializing in work-related injuries to pursue his passion for photography.

Largely self-taught, David studies technique, style, composition, color theory, digital editing software, and printing. He has freelanced extensively, specializing in magazine foodwork and promotional work for non-profits.

David enjoys using photography as a vehicle to help others showcase their good work and has a strong interest in the environment, particularly all creatures great and small. He works for world-renowned painter Ulla Darni on the reproduction side of her business using the digital skills he has acquired. He can be reached at david@davidsaxphoto.com.

Photographer **Jill Meyers** grew up in Brooklyn, New York, and was a teacher in New York City public schools for thirty-four years. Much as she loved her profession, she also loves being "retired" and living in the country equally and having time to pursue her many passions. Jill credits Catskill Animal Sanctuary with changing her life.

It has only been since her retirement that Jill turned her passion for photography into a life purpose. Jill has lived in the Hudson Valley since 2006 and has expanded her photographic repertoire to include farm animals, mainly those residing at Catskill Animal Sanctuary, where her focus is on capturing moments between the animals at CAS and their human friends.

Join Us!

No matter whether you're in New York or New Delhi, there's a way to get involved in the work of Catskill Animal Sanctuary. We'd be delighted and grateful to have you join the team! To learn more about membership, sponsorship, volunteering, tours, scheduling me to speak, special events, and more, check out our website: casanctuary.org. You can also reserve your room at the Homestead, sign up for cooking classes, and register your child for Camp Kindness! Meantime, I hope you'll join the conversation on *Huffington Post*. I blog at least monthly, more when I'm able, and love chatting with readers. From all of us at CAS, thanks!

Catskill Animal Sanctuary

316 Old Stage Road
Saugerties, NY 12477
www.casanctuary.org
845-336-8447